The 23rd Cycle

The 23rd Cycle

Learning to Live with a Stormy Star

Sten F. Odenwald

COLUMBIA UNIVERSITY PRESS NEW YORK

Columbia University Press
Publishers Since 1893
New York, Chichester, West Sussex

Library of Congress Cataloging-in-Publication Data

Odenwald, Sten F.
The 23rd cycle : learning to live with a stormy star / Sten F. Odenwald.
p. cm.
Includes bibliographical references and index.
ISBN 0-231-12078-8 (cloth) — ISBN 0-231-12079-6 (pbk.)
1. Solar activity—Environmental aspects. I. Title: Twenty-third cycle.
II. Title.

QB524.O34 2000
538′.746–dc21 00-059647

Casebound editions of Columbia University Press books are printed on permanent and durable acid-free paper.
Printed in the United States of America
c 10 9 8 7 6 5 4 3 2 1
p 10 9 8 7 6 5 4 3 2 1

To my parents, Rosa and Sten, who never got to see their son's handiwork

Contents

Part III, The Future

Acknowledgments

The writing of a book such as this was an exciting process, made even more so by many people who helped me understand the dimensions of this subject and express it clearly. My wife, Sue, and my daughters, Emily Rosa and Stacia Elise, gave me a tremendous amount of support by just "being there" and understanding my peculiar early morning writing rituals. We are all going to look forward to having more breakfasts together again! I would like to thank my editor, Holly Hodder at Columbia University Press, for her enthusiasm for this project and the many excellent suggestions she made in helping me organize this material to make it readable. If you should find this book both thought provoking and captivating, it is largely to Holly's credit. I would also like to thank Susan Pensak for her excellent job of copyediting this book.

A subject as large as this, with as many facets, has to be written with great care, and I am grateful for the help I received from many colleagues and experts in space weather issues, NASA policy, and the industrial community. I would like to thank Joe Allen at the National Geophysical Data Center for his careful reading of the manuscript and numerous excellent suggestions and comments. I also thank George Withbroe, director of NASA's Office of Space Science, for explaining to me NASA's Living with a Star program. Any errors or misunderstandings you may uncover in this book about current policy, budget, or program issues in space weather are entirely the fault of the author. I would like to thank James Burch, Shing Fung, Dennis Gallagher, Jim Green, Pat Reiff, and Bill Taylor of the *IMAGE* satellite project for many conversations about space weather issues and *IMAGE* science objectives. I am grateful to John Kappenman at Metatech for helping me to understand the electrical power industry and GICs. I would like to thank Art Poland, former project manager for the NASA SOHO program, Eric Christian, ACE deputy project scientist, and Tycho von Rosenvinge, ACE coinvestigator, for their insight into how these space science missions operate. Barbara Thompson, E. Stassinopoulos, and Michael Lauriente at the NASA Goddard Space Flight Center were most helpful in explaining to me how individual researchers in space science receive their funding and how radiation mitigation issues are being investigated. Mike Vinter, vice president of International Space Brokers, Inc., was very helpful in describing the way that satellite insurers operate, which for me dispelled several important misconceptions about this fascinating and highly volatile subject.

There was a great deal of material that had to be trimmed to keep the story focused. Please visit the Astronomy Cafe web site (http://www.theastronomycafe.net) and click on the link for Space Weather. This page contains notes for each chapter and more bibliographic information about space weather issues. Updates on the progress of the 23rd Cycle and its impacts can be found on-site as the information becomes available as well. I will also post any corrections to material in the printed version of this book. If you have questions or comments about this book, or on the subject, please visit the web site and send me an e-mail letter. I will post these in a public FAQ area on the web site along with my responses, as time permits.

Prologue

The 23rd Cycle is certainly an odd-sounding title for a book. Chances are, without the subtitle, *Learning to Live with a Stormy Star,* you might think this is a book about a new washing machine setting or some New Age nonsense. Instead, what you are going to find is a story about how we have misjudged what a "garden variety" star can do to us when we aren't paying attention. Consider this: solar storms have caused blackouts that affect millions of people; they have caused billions of dollars of commercial satellites to malfunction and die; they may also have had a hand in causing a gas pipeline rupture that killed five hundred people in 1989. Despite this level of calamity, the odds are very good that you have never heard about most of these impacts, because they are infrequent, the news media does not make the connection between solar storms and technological impacts, and there are powerful constituencies who would just as soon you not hear about these kinds of "anomalies."

For over 150 years, telescopic views of the Sun's surface have revealed a rhythmic rise and fall in the number of sunspots. Each cycle lasts about eleven years from "sunspot maximum" to "sunspot maximum," and, in step with this, scientists have found many other things that keep a rough cadence with it. The Northern and Southern Lights (aurora) are more common during sunspot maximum than minimum. Titanic solar flares brighter than a million hydrogen bombs also come and go with this cycle. But there is a darker side to these events. Solar flares can kill, aurora can cause blackouts, and satellites can literally be forced out of the sky.

My own professional contact with solar activity came in the 1990s when a change in my working circumstances found me confronting the various hobgoblins of "space science" for the first time since graduate school. These kinds of changes are usually a wake-up call for most people, but for me it meant that a fifteen-year research program in infrared astronomy had come to an end. NASA's COBE satellite program ended in 1996, and so, for a variety of complicated reasons, did much of my full-time research. For the first time, I found myself with only enough grant money to support my career as an astronomer for eight hours a week. In my case, the Sun's talent for raising havoc became something of a professional life preserver.

Very luckily, NASA had just given the go-ahead to James Burch at Southwest Research Institute in San Antonio, Texas, to begin work on the *Imager for Magenetopause-to-Auroral Global Exploration* (*IMAGE*). It was a satellite that would orbit the Earth and keep watch on the movement of energetic particles as the Sun "threw its various tantrums." Although they didn't have much use for an astronomer, they did have funds to set up an education and public outreach program. This program would be handled by Raytheon's Information and Technical Services, Maryland division—my employer. It didn't take long before William Taylor, who was the director of the *IMAGE* education and outreach effort, hired me to help turn their proposed program, called POETRY (Public Outreach, Education, Teaching, and Reaching Youth), into a real flesh-and-blood education program for students, teachers, and the general public.

I began to realize that space science was a very long way from the kind of astronomical research I had been doing for the last fifteen years. I was unfamiliar with the field's scientific issues, and I had hardly a clue about how to capture the public's imagination in an area I regarded as rather far removed from the public's mind. It had nothing to do with gravity, black holes, cosmology, or the topography of the Milky Way. It had everything to do with magnetism, the Sun, and invisible processes operating around the Earth.

And now I have a confession to make.

Hardly any astronomer I know really enjoys space physics of the kind involved in studying the Sun-Earth system. Before the Space Age, space science was an area of research not many young astronomers found much stimulation in. The excitement of exploring how stars evolve, and the structure and contents of the universe, was a much more potent draw of attention and enthusiasm. Solar and space research was often seen as too local, and it was intellectually very messy physics, to boot. In these areas of physical science, the simple relationships and mathematical formulations of Isaac Newton's universal gravitation were almost irrelevant. The particles and winds that blow from the Sun are a charged plasma that drag with them magnetic fields. The geospace environment is another system of plasma and magnetic fields distinct from the Sun but nevertheless electrically connected to it by the solar wind. The relevant principles in physics that have to be mastered are not those of Newton's gravity. Instead, it is James Clerk Maxwell's electrodynamics that take center stage. Currents and fields coexist in complex equations sprouting curlicue letter ∂'s and inverted triangles ∇—the machinery of vector differential calculus. Because plasmas contain charged particles, they interact through electromagnetic forces trillions of times stronger than gravitational ones. Clumps of plasma in one part of the system can interact with other remote clumps and produce complex collective interactions and patterns of motion. The currents spawned by these motions generate their own magnetic fields, which can modify already existing ones in distant corners of the system. Very ugly stuff to the average astronomer! Because of this professional bias within the astronomical community, you probably know more about the subtleties of Big Bang cosmology, whose key event happened *fifteen* billion years ago, or Europa's subsurface sea, than you do about what the Sun is doing right now. The irony is, however, that while you will never have to worry about quasars and supernovae ruining your day, you may have cause to worry about the next big solar flare or ejection of solar plasma!

In the middle of trying to master decades of research in unfamiliar corners of space physics, I made a remarkable personal discovery. Here and there, I found mixed in with the physics brief references to the impacts that these processes have on our technology and ourselves. Blackouts? Satellite malfunctions? Radio interference? What was all this stuff?

Astronomers have always worked in an arena in which virtually all of what we study has zero impact on individual human lives. The closest astronomers ever come to having a direct human impact is when we explain the lunar and solar tides, which are the blessing of surfboarders around the world, or the constancy of the Sun's light and heat. When we discussed astronomical research with the general public, we wrote about black holes and the Big Bang, investing them with awe and wonderment. But we knew full well that this was about as far as we could go in touching upon the practical benefits of research. Fortunately, the general public also values these insights, and, like astronomers, they find the exploration of space an endlessly fascinating story. So all is well.

But now my perspective has changed. What I discovered (and what space scientists had never forgotten in the first place) was that the Sun gives us far more than just a lovely sunny afternoon. Something called "solar storms" can leap out from the Sun and unleash a cascade of events from one end of the solar system to the other. Reaching Earth, some of them even make intimate contact with everything from the light switch on your wall to that pager or cellular phone you carry in your pocket. They can paint the sky with dazzling color, plunge millions of people into darkness, or rob them of their freedom to communicate.

Here, amongst the complex calculus of plasma physics, I came into contact with a dramatic world of things moving in darkness, of human impacts, of calamity. For the first time in my professional life, sterile equations in astrophysics came alive with measurable human consequences. A flow of particles in one place could "toast" a satellite and silence over forty million pagers. A similar current elsewhere could cause an ephemeral aurora to dance in the sky and make you gasp in wonderment, and make you feel that something divine was taking place.

So where was the literature on all these impacts? Why had I never heard about this before in all my daily readings about frontier science? The reason is that it was tucked away among countless anecdotes, papers, books, and newspapers like filler, serving only to enliven long expositions on the underlying physics of aurora or solar physics. Much of it was also out-of-date and hackneyed as author after author rehashed the same three or four spectacular incidents. Yet I had never heard of any of these examples of astrophysics made personal, and each one was uncovered like a diamond sifted from the river silt. Very soon, though, I had accumulated a bucketful of these diamonds, and it was now time to make sense of what I had found. The human impacts were not scattered events in space and time; they were a legacy, written in our very technology, of work left undone, and problems endlessly repeated, that have dogged us for centuries. Hearing about these incidents was like hearing for the first time about tornadoes and then trying to collect reports of their various comings and goings.

Eventually, as I moved among researchers in space science, I also began to encounter a most curious undercurrent of hushed comments and anecdotes that seemed just a trifle too melodramatic. Could it *really* be true that satellite manufacturers didn't want scientists to reveal just how vulnerable their satellites were to solar storms? Was NASA trying to downplay scientific studies of satellites being "killed" by space weather events? Could space-suited astronauts be in more danger for radiation poisoning than anyone wanted to publicly admit? The list seemed endless, and the implications seemed a bit more distressing than anything an astronomer might ever encounter in writing about dark matter or the cosmological constant. Physics and space science seemed to be in bed with the darker side of human foibles in any accounting that described how space physics affects the individual. Would it be possible—or even desirable—to present only the facts, shorn of their implications, both political and economic?

Space weather, as I soon learned it was called, touches on more than just sterile technology. This technology is built by humans for many different commercial and military purposes. With every report of an impact, a protest or denial would be registered, an accusation of ineptitude or intentional wrongdoing would be pronounced. At first, I could see no way out of it. It would not be possible to mention a problem spawned by adverse space weather without giving the impression that the owner of the technology had been asleep at the switch or profoundly naive. It would not be possible to mention human radiation exposure without sounding alarmist or implying between the lines that some governmental agency was negligent in assessing actual health risks.

There is, however, a way to present the human impact of space weather to tell the story and allow it to provide its own interpretation. Like the reactions in the core of a star, the individual components to the story are inert until they are fused together to shed a bit of light on the subject.

We are going to see that the long arm of the Sun can reach deep into many unsuspected niches of our technological civilization, causing blackouts, satellite problems, or pipeline corrosion. Navigation systems that rely on compass bearings can become temporarily confused by "magnetic storms." Short wave signals have been routinely disrupted for hours, rendering long distance communication and LORAN navigation beacons useless or unreliable. Even the atmosphere itself can become our own worst enemy, dragging satellites to a fiery doom.

Each time a major solar flare erupts, the energetic particles that reach the Earth collide with atoms in the atmosphere. The collision liberates high-speed neutrons that can penetrate jet planes, homes, our bodies, and our most advanced technologies. Even the breakneck pace of computer technology development may be restrained by neutron showers as integrated circuit chips become smaller and faster.

So why should we care that we are now once again living under sunspot maximum conditions during Cycle 23? After all, we have already weathered at least five of these solar activity cycles since the end of World War II—nearly a dozen in the twentieth century alone. What is different about the world today is that we are substantially more reliant upon computers and telecommunications to run our commerce and even our forms of entertainment and recreation. The 15 communications satellites we had in 1981 have been joined by 350 in 2000. Cellular phones, PCs, and the Internet have become an overnight $100 billion industry. To support all this, not only will we need more satellites, we will need more electricity flowing in our power grid, which will have to work under loads unheard of in the past. As voters continue to elect *not* to build more power plants, even the North American Electric Reliability Council forecasts that blackouts and brownouts will become more common as power companies run out of temporary sources of power to buy during peak-load conditions in summer and winter.

Although no one can say for sure how current trends are going to play themselves out in the next five to ten years, the evidence that demonstrates the ways we have *already* been affected is well documented. It all comes down to the simple fact that the Sun is not the well-behaved neighbor we would like to imagine it to be. It pummels us every few days or weeks with dramatic storms launched from the surface at millions of miles per hour. Between the solar surface and the Earth's surface, all our technology and human activity plays itself out as if between the proverbial rock and hard place. In most cases, we can not even tell when the next blow is likely to fall. But there is no great mystery about what is going on. We have had a long history—spanning a century or more—of calamities spawned by solar disturbances. It is from this record that we can begin to see what problems may be lurking just around the corner. As the Sun continues to cycle up and down—some twenty-two times since the mid-1700s—the confluence of technological innovation and human commercial necessity now finds us at greater risk for trouble during this, the 23rd Solar Cycle, than in many previous ones. What has changed is the level of our reliance upon sophisticated technology and its widespread infiltration into every niche of modern society. What has not changed is our possibly misplaced sense of confidence that this too will pass with no real and lasting hardship. The issue is not who is responsible for today's suite of vulnerabilities, but what they are preparing to do about them from this moment onward.

Part I

The Past

1 A Conflagration of Storms

All those motorists sitting at traffic lights cursing, should realize that it is not Hydro-Quebec's fault.

—Hydro-Quebec, 1989

On Thursday, March 9, 1989, astronomers at the Kitt Peak Solar Observatory spotted a major solar flare in progress. Eight minutes later, the Earth's outer atmosphere was struck by a blast of powerful ultraviolet and X-ray radiation. The next day, an even more powerful eruption launched a cloud of gas thirty-six times the size of the Earth from Active Region 5395 nearly dead center on the Sun. The storm cloud rushed out from the Sun at over one million miles an hour, and on the evening of Monday, March 13, it struck the Earth. Alaskan and Scandinavian observers were treated to a spectacular auroral display that night. Intense colors from the rare Great Aurora painted the skies around the world in vivid shapes that moved like legendary dragons. Ghostly celestial armies once again battled from sunset to sunrise. Newspapers that reported this event considered the aurora itself to be the most newsworthy aspect of the storm. Viewed as far south as Florida, Cuba, and Mexico, the vast majority of people in the Northern Hemisphere had never seen such a spectacle. Some even worried that a nuclear first strike might be in progress.

Luke Pontin, a charter boat operator in the Florida Keys, described the colors as iridescent reddish hues when they reflected from the warm Caribbean waters. In Salt Lake City, Raymond Niesporek nearly lost his fish while staring transfixed at the northern display. He had no idea what it was until he returned home and heard about the rare aurora over Utah from the evening news. Although most of the Midwest was clouded over, in Austin, Texas, meteorologist Rich Knight at KXAN had to deal with hundreds of callers asking about what they were seeing. The first thing on many people's minds was the Space Shuttle *Discovery* (STS-29), which had been launched on March 13 at 9:57:00 A.M. Had it exploded? Was it coming apart and raining down over the Earth? Millions marveled at the beautiful celestial spectacle, and solar physicists delighted in the new data it brought to them, but many more were not so happy.

Silently, the storm had impacted the magnetic field of the Earth and caused a powerful jet stream of current to flow sixty miles above the ground. Like a drunken serpent, its coils gyrated and swooped downward in latitude, deep into North America. As midnight came and went, invisible electromagnetic forces were staging their own pitched battle in a vast arena bounded by the sky above and the rocky subterranean reaches of the Earth. A river of charged particles and electrons in the ionosphere flowed from west to east, inducing powerful electrical currents in the ground that surged into many natural nooks and crannies. There, beneath the surface, natural rock resistance murdered them quietly in the night. Nature has its own effective defenses for these currents, but human technology was not so fortunate on this particular night. The currents eventually found harbor in the electrical systems of Great Britain, Scandinavia, the United States, and Canada.

At 2:44:16 A.M. on March 13, all was well, and power engineers at Hydro-Quebec resigned themselves to yet another night of watching loads come and go during the off-peak hours. The rest of the world had finished enjoying the dance of the aurora borealis and were slumbering peacefully, preparing for another day's work. The engineers didn't know, however, that for the last half-hour their entire system had been under attack by powerful subterranean Earth currents. One second later, at 2:44:17 A.M., these currents found a weak spot in the power grid of the Hydro-Quebec Power Authority. Static Volt-Ampere Reactive (VAR) capacitor Number 12 at the Chibougamau substation tripped and went offline as harmonic currents induced by the electrojet flowing overhead caused protective relays for this 100-ton behemoth to sense overload conditions. In its wake, the loss of voltage regulation at Chibougamau created power swings and a reduction of power generation in the 735,000-volt La Grande transmission network. At 2:44:19 A.M., at the same station, a second capacitor followed suit. Then, 150 kilometers away, at the Albanel and Nemiskau stations, four more capacitors went off-line at 2:44:46. The last to fall, at 2:45:16 A.M., was a capacitor at the Laverendrye complex to the south of Chibougamau. The fate of the network had been sealed in barely fifty-nine seconds as the entire 9,460-megawatt output from Hydro-Quebec's La Grande Hydroelectric Complex found itself without proper regulation.

In less than a minute, Quebec had lost half its electrical power generation. Automatic load-reduction systems tried to restore a balance between the loads connected to the power grid and the massive loss of capacity now available. One by one, load-reduction systems disconnected towns and regions across Quebec, but to no avail. Domestic heating and lighting systems began to flicker and go out as the emergency load-shedding operation continued its desperate cascade. Eight seconds later, at 2:45:24 A.M., power swings tripped the supply lines from the 2,200 megawatt Churchill Falls generation complex. By 2:45:32 A.M., the entire Quebec power grid collapsed, and most of the province found itself without power. The domino fall of events was much too fast for human operators to react, but it was more than enough time for 21,500 megawatts of badly needed electrical power to suddenly disappear from service.

The nighttime temperature in Toronto was 19 degrees F (−6.8 C), with a high temperature that day of only 34 F (1.6 C), so the loss of electrical power was felt dramatically when most people woke up to cold homes for breakfast. Over three million people live near Montreal, the second largest metropolitan area in Canada, where nearly half of the population of Quebec resides. It is famous for its 30 kilometers of underground walkways linking sixty buildings, two universities, and thousands of shops and businesses. Over five hundred thousand people use this system each day to avoid the bracing cold winter air. Pedestrians using this electrically lit system suddenly found themselves plunged into complete darkness, with only the feeble battery-powered safety lights to guide them to the surface.

The presses at the *Montreal Gazette* had been rolling at breakneck speed that night to print the Monday newspaper for its 195,000 subscribers, but the power failure shut the production down for a day. One can imagine huge rolls of paper, weighing several tons each, coming to a sudden halt, shredding in a storm of debris and jamming the presses. The *Montreal Gazette* apologized to its customers for the undelivered morning paper and blamed what they had assumed was a local power failure in Montreal. Their sister newspaper, *La Presse,* meanwhile, seemed unaffected by the outage and was more than happy to help the *Gazette* press their papers. The only casualty was the color comics section, which came out a day later. Dealing with their own emergency, they had little time to investigate just what had happened. A cursory call to Hydro-Quebec identified the cause of the outage as a defective 12,000-volt cable that provided the *Gazette* with power. There was no mention of any aurora sighted in Montreal, because this display was now gracing the skies of cities hundreds of miles to their south. The five thousand subscribers who called the newspaper that day didn't want to hear about the blackout. They just wanted their morning paper delivered. On March 14, the tone of reportage changed rather abruptly, when the details of what had actually happened were finally put together. It turned out to be quite a story.

The blackout closed schools and businesses, kept the Montreal Metro shut down during the morning rush hour, and paralyzed Dorval Airport, delaying flights. Without their navigation radar, no flight could land or take off until power had been restored. People ate their cold breakfasts in the dark and left for work. They soon found themselves stuck in traffic that attempted to navigate darkened intersections without any streetlights or traffic control systems operating. Like most modern cities, people work round the clock, and in the early morning hours of March 13, the swing shift staffed many office buildings in the caverns of downtown Montreal. All these buildings were now pitch dark, stranding workers in offices, stairwells, and elevators. By some accounts, the blackout cost businesses tens of millions of dollars as it stalled production, idled workers, and spoiled products.

Hydro-Quebec officials insisted that the vast power system was, itself, innocent of blame. The fault, they said, was in the geography of Quebec, which had power lines extending much farther north than other electrical systems. Many people soon pointed out that this was the second major blackout in less than a year, and that, when you added up the numbers, Hydro-Quebec's outages totaled about nine hours per year compared to neighboring Manitoba Power and Electric's two-hours-per-year average. Hydro-Quebec promised to invest another $2 billion to cut in half the number of yearly blackouts, but this didn't derail the investigations that were called for by the government to see if Hydro-Quebec had been negligent. Energy Minister John Ciaccia echoed the sentiments of many people as they sat in snarled traffic facing blackened signals, "It's frustrating because, despite all our efforts to upgrade the system, we still wake up at 5 A.M. with a total blackout."

By 10:00 A.M., power had been restored to most of the customers in Quebec. An hour later, all but 3,500 of the 842,000 customers were back in business. Picking up the pieces was not going to be easy in a system as large as Hydro-Quebec, with its thousands of miles of power lines and hundreds of transformers and other electrical components. It all had to work perfectly or another blackout would result. Isolated power failures were promised over the next twenty-four hours as Hydro-Quebec wrestled with carefully restarting their vast interconnection of power lines and transformers. Residential customers, they announced, would be at the bottom of the priority list for reconnection.

New York Power authorities lost 150 megawatts the moment Hydro-Quebec went down, and the New England Power Pool lost 1,410 megawatts at about the same time. Service to ninety-six utilities in six New England states was interrupted while other reserves of electrical power were bought and brought on-line. In a show of solidarity with their sister utility in the North, by 9:00 A.M., New York Power and NEPool were sending over 1,100 megawatts of power to Quebec to tide them over while the system was being brought back up again. Luckily, these states had the power to spare at the time. But just barely. Some of them had their own cliff-hanger problems to deal with. Electrical power pools serving the northeastern United States had come very close to going down as well.

The intense electrojet currents that had flowed in the upper atmosphere had, in a matter of seconds, spread their impact far and wide, causing electrical disturbances throughout North America and Great Britain. A thousand miles away from Hydro-Quebec, Allegheny Power, which serves Maryland, Virginia, and Pennsylvania, lost ten of its twenty-four VAR capacitors as they were automatically taken off-line to avoid damage. A $12 million, 22,000-volt generator step-up transformer, owned by the Public Service Electric and Gas Company of New Jersey, experienced overheating and permanent insulation damage. This transformer was the linchpin in taking electricity from the Salem Nuclear Plant and boosting it to 500,000 volts for long distance transmission. Replacement power had to be bought for $400,000 per day to keep East Coast residents from sharing the same fate as their neighbors in Quebec. Luckily, the owners had a spare replacement transformer available, but it still took six months to install. Without the replacement, it would have taken a year to order a new one. Across the United States, from coast to coast, over two hundred transformer and relay problems erupted within minutes of the start of the storm. Fifty million people in the United States went about their business or slept, never suspecting their electrical systems had been driven to the edge of disaster. Not since the Great Blackout of 1965 had U.S. citizens been involved in a similar outage. There would have been no place they could drive to escape the enfolding darkness had it come at night.

According to Joe Allen, solar and terrestrial physics chief at the National Geophysical Data Center (NGDC), but now retired, the solar flare and accompanying storm conditions did much more than cause a blackout and upset communications systems. Automatic garage doors in California suburbs began to open and close without apparent reason as offshore navy vessels switched to low frequency backup transmitters. Microchip production in the northeastern United States came to a halt several times because of the ionosphere's magnetic activity. In space, geostationary communications satellites that sensed the Earth's magnetic field in order to point themselves had to be manually repointed from the ground as the local field polarity reversed direction, causing satellites to try and flip upside down. Some satellites in polar orbits actually tumbled out of control for several hours. GOES weather satellite communications were interrupted, causing weather images, used by the National Weather Service for their daily forecasts, to be lost. NASA's TDRSS-1 communication satellite recorded over 250 electrical and communications incidents caused by the increased particles as they flowed deep into the satellite's sensitive electronics.

The *Chicago Tribune* and the *Washington Post* said nothing about the storm—or even the blackout for that matter. Only a brief mention was made about it in European papers such as the *London Times*, and then only to comment on the spectacular aurora. The *Fairbanks Daily News* and the *Anchorage Daily News* ran several articles describing the auroral display but also failed to mention the power outage. The *Toronto Star* in Quebec, at least on page 3, considered the blackout in its own province to be a significant news event, and on March 13, 1989, announced, "Huge Storms on Sun Linked to Blackout That Crippled Quebec":

> Fiery storms on the Sun may have caused yesterday's huge power blackout that left almost 6 million people without heat or electricity

for almost 9 hours. . . . Premier Robert Bourassa did not believe the blackout will dissuade U.S. utilities from signing lucrative contracts to buy Quebec electricity, the cornerstone of the premier's economic policies. . . . An official from the New York Power Authority from which Hydro-Quebec bought 700 megawatts, said in an interview he would prefer that Quebec didn't have so many power blackouts.

Meanwhile, the Space Shuttle *Discovery* was having its own mysterious problems. A sensor on one of the tanks supplying hydrogen to a fuel cell was showing unusually high pressure readings on March 13. "The hydrogen is exhibiting a pressure signature that we haven't ever seen before," said the flight director, Granville Pennington, at the Johnson

FIGURE 1.1 Nighttime view of North America showing aurora and city lights.
source: *Defense Meteorological Satellite Program*

Space Center. Engineers tried, apparently unsuccessfully, to understand the odd readings in order to advise whether to end the flight a day early on Friday. No public connection was ever made between this instrument reading "glitch" and the solar storm that crippled Quebec, but it is fair to say that the conjunction of these two events was not completely by chance.

In many ways, the Quebec blackout was a sanitized calamity. It was wrapped in a diversion of beautiful colors and affected a distant population mostly while they slept. There were no houses torn asunder or streets flooded in the manner of a hurricane or tornado. There was no dramatic footage of waves crashing against the beach. There were no cyclonic whirlwinds cutting a swath of destruction through Kansas trailer parks. The calamity passed without mention in the major metropolitan newspapers, yet six million people were affected as they woke to find no electricity to see them through a cold Quebec wintry morning. Engineers from the major North American power companies were not so blasé about what some would later conclude could easily have escalated into a $6 billion catastrophe affecting most U.S. East Coast cities. All that prevented fifty million more people in the U.S. from joining their Canadian friends in the dark were a dozen or so heroic capacitors on the Allegheny Power Network.

Today the March 1989 Quebec Blackout has reached legendary stature, at least among electrical engineers and space scientists, as an example of how solar storms can adversely affect us. It has even begun to appear in science textbooks. Fortunately, storms as powerful as this are rare. It takes quite a solar wallop to cause anything like the conditions leading up to a Quebec-style blackout. When might we expect the next one to happen? About once every ten years or so, but the exact time is largely a game of chance.

Why should we care that we are now once again living under sunspot maximum conditions? After all, we have already weathered at least five of these solar activity cycles since the end of World War II. What is different about the world today is that we are substantially more reliant upon computers and telecommunications to run our commerce, and even our forms of entertainment and recreation. In 1981, at the peak of Solar Cycle 21, there were 15 communication satellites in orbit. Cellular phones were rare, and there were 800,000 PCs sold in the U.S., with 300 hosts on the Internet. By the time the peak of Solar Cycle 22 came around in 1989, there were 102 communication satellites and 3 million cellular phone users in the United States. With the new Intel 80486-based PCs, you could send e-mail to your choice of 300,000 host machines on the Internet.

As we arrive at the peak of the 23rd Sunspot Cycle in 2000–2001, however, we enter a very different world, far more reliant on what used to be the luxuries of the Space Age. By 2000, 349 communication satellites orbit the Earth, supporting over $60 billion of commerce. Over 100 million people have cellular phones, and Global Positioning System (GPS) handsets are a commonplace for people working, or camping, "off road." By 2003, 400 million people will routinely use wireless data transmission via satellite channels. There will be over 10 million Internet hosts, with 38 percent of U.S. households Internet-connected. To support all this, not only will we need more satellites, but we will need more electricity flowing in our power grid, which will have to work under loads unheard of in the past. As voters continue to elect not to build more power plants, blackouts and brownouts will become more common as power companies run out of temporary sources of power to buy during peak-load conditions in the summer and winter.

As if to emphasize today's exuberance and expectations, *Individual Investor* magazine announced, on its cover, "The Sky's the Limit: In the 21st Century Satellites Will Connect the Globe." The International Telecommunications Union in Geneva has predicted that by 2005 the demand for voice and data transmission services will increase to $1.2 trillion. The fraction carried by satellite services will reach a staggering $80 billion.

To meet this demand, many commercial companies are launching not just individual satellites but entire networks of them, with names like *Iridium*, *Teledesic*, *Skybridge*, and *Spaceway*. The total cost of these systems alone represents a hardware investment of $35 billion between 1998 and 2004. The actual degree of vulnerability of these systems to solar storms is largely unknown and will probably vary in a complex way, depending on the kind of technology they use and their deployment in space. They do, however, share some disturbing characteristics: they are all light weight, sophisticated, built at the lowest cost, and following only a handful of design types replicated dozens and even hundreds of times, often with off-the-shelf electronics.

It is common to base future expectations on recent past experiences: "Past is prologue," some say. Increasingly, these past experiences with, for example, commercial space technology do not extend back much beyond the last solar maximum in 1989–1990. So, when we wonder why infrequent events such as solar storms aren't more noticeable, we have to remind ourselves that most of our experience comes from times when the Sun was simply not very active, and when we were a lot less technologically vulnerable.

Now more than ever, we depend on uninterrupted sources of power. Blackouts are amusing for about the first sixty seconds, then become intolerable. Along with our expensive personal computers, we routinely purchase surge protectors to handle the many intermittent rises and falls of an increasingly complex power delivery system. In the end, no surge protectors can save us from Quebec-style blackouts. We have become dependent on our cell phones and pagers in a way that will tie critical moments in our private lives to the shotgun physics of satellite and power grid survival during invisible solar storms. When a single failed satellite like the *Galaxy IV* in May 1998 can catch over forty million pagers off guard, do we find ourselves more secure? Sometimes it can be dangerous and costly to gamble, although most of the time we seem to get by hardly realizing that a calamity has passed over us. We actually seem to enjoy living on the technological "edge" as we use our cellular phones while driving our cars at 65 mph. But when a corroded natural gas pipeline in the Urals sprung a leak and detonated in June 1989, five hundred people died. Pipelines corrode, and solar storms can hasten this process, with tragic consequences.

Beyond the reality of our unfamiliarity with the connection between solar events and terrestrial difficulties, there is also a disturbing tendency among some communities to deny that there is a serious problem. In both the electrical power industry and in the satellite business, there seems to be a desire not to recognize that certain ventures are inherently risky and intrinsically susceptible to solar and geophysical influences. At the same time that the emplacement of vital communications systems, and human activities in space, has escalated, our scientific understanding of how the Sun affects us has not kept up because of a lack of consistent research funding and badly needed data.

Although no one can say for sure how current trends in thinking are going to play themselves out in the next five to ten years, the evidence for how we have *already* been affected in the past is well documented. It all comes down to the simple fact that the Sun is not the polite and well-behaved neighbor we would like to imagine it to be. Not only do we find ourselves between a rock and a hard place, but we can't even tell when the next blow is likely to fall. Although the timing of the next outage is unpredictable, there is no great mystery about what is going on. We have had a long history—spanning over a century—of calamities spawned by solar disturbances, and in this legacy we can see many forewarnings.

In the chapters to follow, we are going to see why most experts feel we will be at greater risk for trouble during this, the 23rd Solar Cycle, than in many previous ones. What has changed during the last ten years is the level of our reliance upon sophisticated technology and its widespread infiltration into every niche of modern society.

2 Dancing in the Light

He knew, by streamers that shot so bright,
That spirits were riding the northern light.
—Sir Walter Scott, "The Lay of the Last Minstrel," 1802

January 7, 1997, seemed to be an ordinary day on the Sun. Photographs taken at the Mauna Kea Solar Observatory showed nothing unusual. In fact, to the eye and other visible wavelength instruments, the images showed not so much as a single sunspot. But X-ray photographs taken by the *Yohkoh* satellite revealed some serious trouble brewing. High above the solar surface, in the tenuous atmosphere of the Sun, invisible lines of magnetic force, like taut rubber bands, were coming undone within a cloud of heated gas. Balanced like a pencil on its point, it neither rose nor fell as magnetic forces levitated the billion-ton cloud high above the surface. Then, without much warning, powerful magnetic fields lost their anchoring and snapped into new shapes; the precarious balance between gravity and gas pressure lost.

The massive cloud launched from the Sun crossed the orbit of Mercury in less than a day. By Wednesday it had passed Venus: an expanding cloud over thirty million miles deep, spanning the space within much of the inner solar system between the Earth and Sun. At a distance of one million miles from the Earth, the leading edge of the invisible cloud finally made contact with NASA's *WIND* satellite at 8:00 P.M. EST on January 9. By 11:30 P.M. the particle and field monitors onboard NASA's earth-orbiting *POLAR* and *GEOTAIL* satellites told their own stories about the blast of energetic particles now sweeping through this corner of the solar system. Interplanetary voyagers would never have suspected the conflagration that had just swept over them. The cloud had a density hardly more than the best laboratory vacuums.

The tangle of fields and plasma slammed into the Earth's own magnetic domain like some enormous sledgehammer as a small part of the fifty-million-mile-wide cloud brushed by the Earth. Nearly a trillion cubic miles of space were now involved in a pitched battle between particles and fields, shaking the Earth's magnetic field for over twenty-four hours. The storminess in space rode the tendrils of the Earth's field all the way down to the ground in a barrage of activity. Major aurora blazed forth in Siberia, Alaska, and across much of Canada during this long winter's night.

The initial blast from the cloud (astronomers call it a "coronal mass ejection" or CME) compressed the magnetosphere and drove it inside the orbits of geosynchronous communications satellites suspended above the daytime hemisphere, amplifying trapped particles to high energies. Dozens of satellites, positioned at fixed longitudes along the Earth's equator like beads on a necklace, alternately entered and exited the full bore of the solar wind every twenty-four hours as they passed outside the Earth's magnetic shield. Plasma analyzers developed by Los Alamos Laboratories, and piggybacked on several geosynchronous satellites, recorded voltages as high as 1,000 volts, as static electric charges danced on their outer surfaces. It was turning out to be not a very pleasant environment for these high-tech islands of silicon and aluminum.

High-speed particles from the cloud seeped down into the northern and southern polar regions, steadily losing their energy as they collided with the thickening blanket of atmosphere. On January 9, at 8:00 P.M. EST, the darkened but cloudy northern hemisphere skies over Alaska and Canada were awash in a diffuse auroral glow of crimson and green that subtly flowed across the sky as the solar storm crashed against the Earth's magnetic shield. This quiescent phase of activity was soon replaced by a far more dramatic one whose cause is a completely separate set of conditions and events that play themselves out in the distant "magnetotail" of the Earth.

Like some great comet with the Earth at its head, magnetic tendrils trail millions of miles behind it above the nighttime hemisphere. At nearly the distance of the Moon, the Earth's field contorts into a new shape in an attempt to relieve some of the stresses built up from the storm cloud's passage. The fields silently rearrange themselves across millions of cubic miles of space. Currents of particles trapped in the shape-shifting field accelerate as magnetic energies are exchanged for pure speed in a headlong kinetic onrush. Some of the particles form an equatorial ring of current, while others flow along polar field lines. Within minutes, beams of particles enter the Earth's polar regions around local midnight, triggering a brilliant aurora witnessed by residents of northern latitudes in Canada and Europe. The quiet diffuse aurora that Alaskan and Canadian observers had seen during the first part of the evening on January 9 were abruptly replaced by a major auroral storm that lasted through the rest of this long winter's night.

As the solar cloud thundered invisibly by, a trace of the frigid atmosphere was imperceptibly sucked high into space in a plume of oxygen

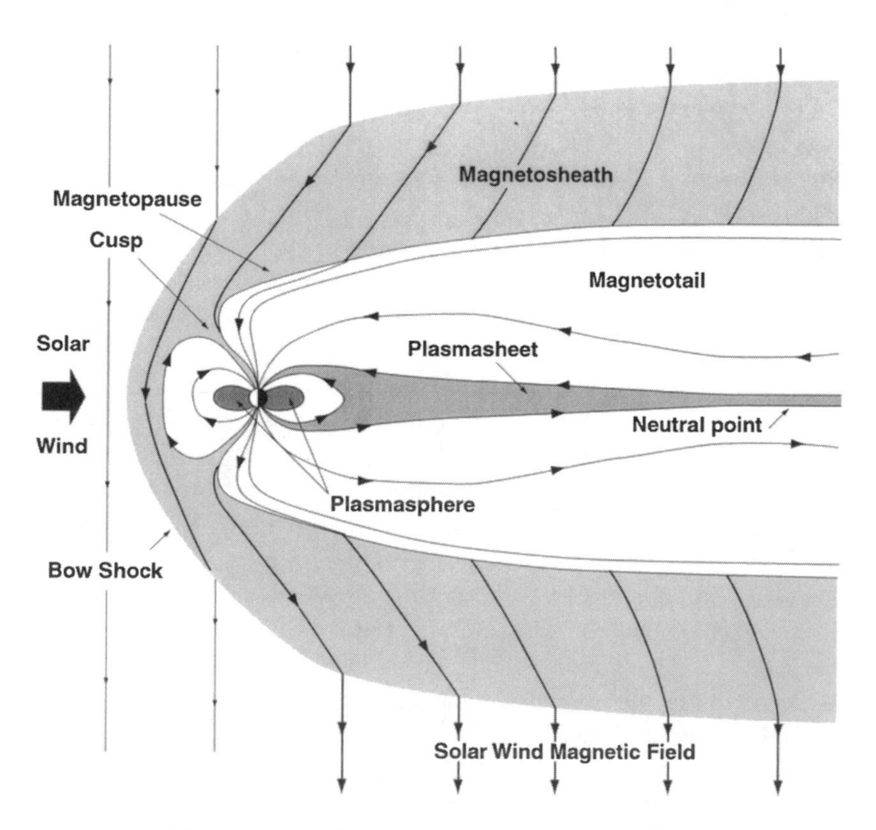

FIGURE 2.1 The major regions of the Earth's space environment showing its magnetic field, plasma clouds, and currents.

and nitrogen atoms. The changing pressure in the bubble wall pumped this fountain as though it were water being siphoned from a well. Atoms once firmly a part of the stratosphere now found themselves propelled upward and accelerated, only to be dumped minutes later into the vast circumterrestrial zone girdling the Earth like a doughnut. Still other currents began to amplify and flow in this equatorial zone. A river of charged particles five thousand miles wide asserted itself as the bubble wall continued to pass. Millions of amperes of current swirled around the Earth in search of some elusive resting place just beyond the next horizon. Like electricity in a wire, this invisible current created its own powerful magnetic field in its moment-by-moment changes as current begets field and field begets more current.

The Earth didn't tolerate the new interloper very well. The current grew stronger, and the Earth's own field was forced to readjust. On the ground, this silent battle was marked by a lessening of the Earth's own field. Compass needles bowed downward in silent assent to magnetic forces waging a pitched battle hundreds of miles above the surface. The same magnetic disturbance that made compasses lose their bearings also infiltrated any long wires splayed out on telegraph poles, in submarine cables, or even in electrical power lines. As the field swept across hundreds of miles of wire and pipe, currents of electrons began to flow, corroding pipelines over time and making messages unintelligible. During the January 1997 storm, the British Antarctic Survey at its South Pole Halley Research Station reported that the storm disrupted high-frequency radio communications and shut down its life-critical aircraft operations.

The storm conditions continued to rage throughout all of January 10, but, just as the conditions began to subside on January 11, the Earth was hit by a huge pressure pulse as the trailing edge of the cloud finally passed by. The arrival and departure of this cloud would not have been of more than scientific interest had it not also incapacitated the $200 million *Telstar 401* communications satellite in its wake at 06:15 EST. The storm had now exited the sphere of scientific interest and landed firmly inside the wider arena of human day-to-day life among millions of TV viewers.

AT&T tried to restore satellite operations for several more days, but on January 17 they finally admitted defeat and decommissioned the satellite. All TV programs such as *Oprah Winfrey*, *Baywatch*, *The Simpsons*, and feeds for ABC News had to be switched to a spare satellite: *Telstar 402R*. The *Orlando Sentinal* on January 12 was the first newspaper to mention the outage in a short seventy-four-line note on page 22. Three days later, the *Los Angeles Times* described how this outage had affected a $712 million sale of AT&T's Skynet telecommunications resources to Loral Space and Communications Ltd. No papers actually mentioned a connection to solar storms until several weeks later, on January 23, when the focus of news reports in the major newspapers was the thrilling scientific studies of this "magnetic cloud." The *New York Times* closed their short article on the cloud by mentioning that "scientists said they do not know if this month's event caused the failure, early on January 11, of AT&T's *Telstar 401* communications satellite, but it occurred during magnetic storms above the earth."

Whether stories get covered in the news media or not is often a matter of luck when it comes to science, and this time there was much that urgently demanded attention. A devastating earthquake had struck Mexico City at 2:30 P.M. on January 10 and cost thousands of lives. This had come close behind the three million people in eastern Canada who had lost power a few days before the satellite outage. In Montreal, over one million people were still waiting for the lights to go back on under cloudy skies and subfreezing temperatures. These were very potent human-interest stories, leaving little room at the time for stories of technological problems caused by distant solar storms. Although the news media barely mentioned the satellite outage, the outfall from this satellite loss reverberated in trade journals for several years afterward, extending far beyond the inconveniences experienced by millions of TV viewers. It was one of the most heavily studied events in space science history, with no fewer than twenty research satellites and dozens of ground observatories measuring its every twist and turn. Still, despite the massive scientific scrutiny of the conditions surrounding the loss of the *Telstar 401* satellite, the five-month-long investigation by the satellite owner begged to differ with the growing impression of solar storm damage rapidly being taken as gospel by just about everyone else.

Although they had no "body" to autopsy, AT&T felt very confident that the cause of the outage involved one of three possibilities: an outright manufacturing error involving an overtightened meter shunt, a frayed bus bar made of tin, or bad Teflon insulation in the satellite's wiring. Case closed. Firmly ruled out was the solar disturbance that everyone seemed to have pointed to as the probable contributing factor. In fact, AT&T would not so much as acknowledge there were any adverse space weather conditions present at all. So far as they were concerned, publicly, the space environment was irrelevant to their satellite's health. In some sense, the environment was, in terrestrial terms, equivalent to a nice sunny day and not the immediate aftermath of a major lightning storm or tornado.

The reluctance of AT&T to make the solar storm–satellite connection did not stop others from drawing in the lines between the dots and forming their own opinions. In some quarters of the news media, the solar storm connection was trumpeted as the obvious cause of the satellite's malfunction in reports such as Cable Network News's "Sun Ejection Killed TV Satellite." *Aviation Week and Space Technology* magazine, a much read and respected space news resource, announced that *Telstar 401*

> suffered a massive power failure on Jan. 11, rendering it completely inoperative. Scientists and investigators believe the anomaly might have been triggered by an isolated but intense magnetic sub-storm, which in turn was caused by a coronal mass ejection . . . spewed from the Sun's atmosphere on Jan. 6.

Even the United States Geological Survey and the United States Department of the Interior released an official fact sheet, titled "Reducing the Risk from Geomagnetic Hazards: On the Watch for Geomagnetic Storms," in which they noted, in their litany of human impacts, "In January 1997, a geomagnetic storm severely damaged the U.S. Telstar 401 communication satellite, which was valued at $200 million, and left it inoperable."

By January 12, the bubble wall had finally passed. The electric connection between the Sun and the Earth waned, and the Earth once again found itself at peace with its interplanetary environment. Its field resumed its equilibrium, expanding back out to shield its retinue of artificial satellites. The currents that temporarily flowed and painted the night skies with their color were soon stilled. Meanwhile, man-made lights on another part of the Earth flickered and went out for eleven minutes. In Foxboro, Massachusetts, the New England Patriots and the Jacksonville Jaguars were in the midst of the AFC championship game when at 5:04 P.M. the lights went out at the stadium. A blown fuse had cut power to a transformer in the Foxboro Stadium just as Adam Vinatieri was lining up to try for a twenty-nine-yard field goal. It may have been mere coincidence, but the cause of the "mysterious" fuse malfunction was never identified in terms of more mundane explanations. The story did make for rather awful puns among sports writers in thirty newspapers from as far away as Los Angeles.

A gentle wind from the Sun incessantly wafts across the Earth, and from time to time causes lesser storms to flare up unexpectedly like inclement weather on a humid summer afternoon. This wind of stripped atoms drags with it into space long fingers of magnetism and pulls them into a great pinwheel pattern spanning the entire solar system. The arms are rooted to the solar surface in great coronal holes, which pour plasma out into space like a dozen faucets. Just as magnets have a polarity, so too do these pinwheel-like arms of solar magnetism. North-type and south-type sectors form an endless and changing procession around the path of the Earth's orbit as old coronal holes vanish and new ones open up to take their place. As the Earth travels through these regions, its own polarity can trigger conflicts, spawning minor storms as opposing polarities search for an elusive balance. The cometlike magnetotail region that extends millions of miles behind the Earth trembles with the subtle disturbances brought on by these magnetic imbalances. Minor waves of particles are again launched into the polar regions as the magnetotail waves like a flag in the wind. Lasting only a few hours, magnetic substorms are nothing like the grand daylong spectacles unleashed by million-mile-wide bubble clouds, but they can be the source of high-energy electrons capable of affecting satellites.

The solar surface can also create storms that traverse interplanetary space with the swiftness of light. Near sunspots where the fields are most intense, explosive rearrangements cause spectacular currents to flow. Crisscrossing currents short-circuit themselves in a burst of heat and light called a "solar flare." Within minutes, the conflagration creates streams of radiation that arrive at the Earth in less time than it takes to cook a hamburger. Within a few hours, a blast of energetic particles traveling at one-third the speed of light begins to arrive, and provides unprotected astronauts and satellites with a potentially lethal rain of disruption. In the ionosphere, the stream of X rays strips terrestrial atoms of their electrons. For hours the ionosphere ceases to act like a mirror to shortwave radio signals across the daytime face of the Earth. The ebb and flow of the atomic pyrotechnics on the distant Sun are reflected in the changing clarity of terrestrial radio transmissions. Soon, the conditions that brought about the solar flare-up mysteriously vanish. The terrestrial atmospheric layers slough off their excess charges through chemical recombination, and once again radio signals are free to skip across the globe unhindered.

As dramatic as these events can be as they play themselves out, they do so under the cloak of almost total invisibility. Unlike the clouds that fill our skies on a summer's day, the motions of the solar plasma and the currents flowing near the Earth have much the same substance as a will-o'-the-wisp. All you can see are the endpoints of their travels in the exhalations from the solar surface or in the delicate traceries of an auroral curtain. Only by placing sensitive instruments in space, and waiting for these buoys to record the passing waves of energy, have we begun to see just how one set of events leads to another, like the fall of dominoes. It is easy to understand why thousands of years had to pass before the essential details could be appreciated.

Over most of Europe and North America, let alone the rest of the world, fewer than two nights each year have any traces of aurora, and if you live in an urban setting, with its light pollution, aurora become literally a once-in-a-lifetime experience. In bygone years, even the urban sky was dark enough that the Milky Way could be seen from such odd places as downtown New York City. Every week or so, in some localities, the delicate colors of the aurora would dance in the sky somewhere. Lacking our modern entertainments of TV, radio, and the Internet, previous generations paid far more attention to whatever spectacles nature could conjure up. For the modern urbanite, this level of appreciation for a natural phenomenon has now become as foreign as the backside of the Moon. Most of the residual legacies of fear and dread that aurora may still command have been substantially muted, especially as our scientific comprehension of the natural world has emerged to utterly demystify them. More important, they do no physical harm, and it is by virtue of this specific fact that they have been rendered irrelevant.

We understand very well the calamities that volcanoes, earthquakes, hurricanes, and tornadoes can cause. For much of human history, volcanic eruptions and earthquakes have been civilization's constant companion, with such legendary episodes as Santorini, Etna, and Krakatoa indelibly etched into the history of the Western world. In the Far East, seasonal typhoons and tsunamis produce flooding the likes of which are rarely seen in the West. It wasn't until settlers expanded throughout the Bahamas, the Gulf of Mexico, and the interior of North America that two new phenomena had to be reckoned with: tornadoes and hurricanes.

TABLE 2.1 A Partial List of Major Aurora Reported Since A.D.1

Date	AA*	Ap*	Areas where visible
A.D. 37			Rome
1/15/1192			Flanders
10/5/1591			Nuremberg
2/10/1681			Pressberg, Hungary
3/17/1716			Europe
9/15/1839			London
11/17/1848			California, Cuba, Europe
8/28/1859			Jamaica, Cuba, Rome, London
9/2/1859			Central America, Greece
2/4/1872	134		India, Mexico, Greece
11/18/1882	372		Cuba, Mexico, New York
3/30/1886	186		England, China, Japan, India
2/13/1892	271		Iowa, New York
3/30/1894	103		England
9/9/1898	179		Omaha, Tennessee, New York
11/1/1903	324		France, New York, California
2/9/1907	174		England
9/25/1909	333		Singapore
8/7/1917	146		New York, Chicago
3/22/1920	241		Boston, Washington, Norway
5/14/1921	356		England, Samoa, Jamaica
1/26/1926	151		U.S., England, Scandinavia
1/25/1938	241	146	U.S., Azores, North Africa
3/24/1940	377	277	New England, North Dakota
9/18/1941	429	312	South Carolina, Indiana
3/28/1946	329	215	????
7/26/1946	200	212	New England, Tennessee
8/19/1950	161	203	????
2/24/1956	131	104	Iceland, Alaska
9/4/1957	211	221	Colorado, Kansas, Michigan
9/13/1957	163	160	Mexico
9/22/1957	206	186	Mexico, U.S., British Columbia
2/10/1958	298	199	North America, Mexico, USSR
7/8/1958	314	216	Canada, Kentucky, New England
9/3/1958	185	171	Canada
7/15/1959	357	252	????
3/15/1960	62	56	Pennsylvania, New York, Dakotas
3/31/1960	312	251	New England

TABLE 2.1 A Partial List of Major Aurora Reported Since a.d.1 *(continued)*

Date	AA*	Ap*	Areas where visible
10/6/1960	253	258	Indiana, Virginia, New England, Ontario
5/25/1967	279	241	California, Tennessee, Alberta
3/23/1969	115	117	North America
8/4/1972	290	223	North America, Canada, Scandinavia
3/5/1981	100	82	California, Colorado
3/13/1981	80	50	Arizona
7/13/1982	268	229	????
9/5/1982	207	201	????
2/8/1986	287	228	????
3/13/1989	441	285	North America
10/29/1991	154	156	New England
11/8/1991	235	179	New Jersey, Iowa, Kansas, Mississipppi
5/10/1992	196	193	????
5/4/1998	136	120	????
9/25/1998	236	????	
10/22/1999	207	U.S.	
1/7/2000	300		Florida, Canada, Europe

NOTE: AA* is an index used by space scientists to indicate the auroral activity of the particular storm and has been used since the 1880s. The Ap* index is related to the geomagnetic disturbance Kp index and was introduced in the 1930s to characterize the magnitude of geomagnetic storms. A number of apparently major aurora as indicated by large AA* and Ap* values seem not to have been reported (i.e., photographed) or considered otherwise noteworthy, especially since 1980.

Indices courtesy Joe Allen NOAA/NGDC. Aurora reports since 1940 obtained from archival editions of *Sky and Telescope* and *Astronomy.*

As human populations established themselves in the New World, our vulnerability to the devastating effects of tornadoes and hurricanes increased year after year.

Like the settlers of Kansas who discovered tornadoes for the first time, we have quietly but steadily entered an age in which aurora and solar storms have become more than just a lovely but rare nighttime spectacle. During the last century, new technologies have emerged such as telegraphy, radio, and satellite communication—and all these modern wonders of engineering have shown consistent patterns of vulnerability to these otherwise rare and distant phenomena. The most serious disruptions have followed relentlessly the rise and fall of the sunspot cycle and the appearance of Great Aurora, which, as table 2.1 shows, arrive every ten years or so and are often seen worldwide. Like tornadoes, we have had to reach a certain stage in our colonizing the right niches in order to place ourselves in harm's way.

Aurora are only the most obvious sign that more complex and invisible phenomena are steadily ratcheting up their activity on the distant solar surface and in the outskirts of the atmosphere high above our heads. When great auroral storms are brewing, they inevitably lead to impacts on our technology that can be every bit as troublesome and potentially deadly as an unexpected lightning storm or a tornado.

3 "Hello? Is Anyone There?"

Up, up, up past the Russel Motel
Up, up, up to the Heavyside Layer
—T. S. Elliot, ca. 1937

It was a fantastic aurora–the best that anyone could recall in decades. When the September 18, 1941, Great Aurora took the stage, it was seen in Virginia, Denver, and St. Louis, but in New York City its displays played to a very mixed audience. From Central Park, at 9:30 P.M., pedestrians could plainly see several bright colored bands of light rivaling the full moon and spanning the sky in shades of orange, blue, and green. Curtains, rays, and flashing displays of light covered much of the sky throughout the rest of the night, giving New Yorkers a taste of what their northern relatives in Alaska see on a weekly schedule. The display had started before sunrise on Thursday, September 18, as thousands of commuters got up and had breakfast before dealing with another New York rush hour. This was a special day for other reasons as well. The Brooklyn Dodgers would be playing the Pittsburg Pirates, and Red Barber would be announcing the play-by-play activity over WOR radio. By 4:00 P.M., the baseball teams were tied 0–0 in a game that kept everyone at the edge of their seats, when suddenly, and for an interminable fifteen minutes, the broadcast was cut off by auroral interference. When the broadcast resumed, the Pirates had scored four runs, and Dodger fans pounded the radio station switchboards by the thousands, hurling oaths and bad language. The radio station tried to explain that they were absolutely blameless and that the fans should be cursing the auroral displays over their heads. Apparently, calm reasoning did little good. No one really bought the idea that solar storms had raided the game.

NBC, meanwhile, was busy trying to resolve their own problems. They were scheduled to do a special inaugural broadcast to Mexico to open twenty-three new affiliate stations. Although the program could be easily heard in the United States, in Mexico the auroral static and interference made the reception of the program impossible. Throughout most of Thursday, NBC and CBS shortwave transmissions were badly interrupted just about everywhere. RCA could no longer make connections with London directly, but they discovered that a new channel for their London broadcasts had opened up instead. By transmitting to Buenos Aires and then having the signal relayed to London from there, along a twelve-thousand-mile path, they could get a connection that was actually clearer than what they usually got along the direct route.

The next day, after dazzling aurora had washed the skies the evening before, New Yorkers were treated to a second not so amusing incident. At 11:45 A.M., WAAT in New Jersey was broadcasting recorded songs by Bing Crosby when a conversation between two men interfered with portions of the music. Station engineers worked frantically to clear up the cross-talk problem, but there wasn't a whole lot they could do. Within a few minutes, the voices just as mysteriously disappeared, but not before callers from New Jersey complained by the hundreds on the station's switchboards.

No sooner had this problem solved itself when the noon news broadcast was clobbered by a much louder conversation between two women. This time the discussion was about their blind dates, and the language they used was politely called "spicy" in newspaper accounts of the incident. Again, the auroral conditions overhead had mixed a shortwave channel with the normal broadcast at nearly the same frequency. Many callers complained about the change in programming, which was being heard by young children. There were even a number of men who called WAAT to ask about the women and whether the station were running a dating service.

Because the powerful electromagnetic forces that accompany auroras have a strong affinity for all things electrical, it is not surprising in retrospect that every communications technology we have come up with in the last 160 years has fallen victim to interference from these natural events. In this, our history of understanding aurora is intimately interwoven with the near simultaneous rise in understanding electricity and magnetism.

Electrical currents are actually just as magical as aurora in many of the ways that they work. For example, in 1820, Hans Oerstead, a physicist at the University of Copenhagen, could make electrical currents deflect compass needles. Meanwhile, across the English Channel, Michael Faraday uncovered an equally mysterious magnetic phenomenon: if you move a magnet across a wire, it causes a current to flow in the wire. It's hard to imagine the excitement these investigators must have felt as they saw electrical currents produce invisible magnetic forces and vice versa. Faraday's discovery of changing magnetic fields producing electrical currents, combined with Alexander von Humbolt's discovery that sudden changes in the Earth's magnetism can occur in "magnetic storms," provided the ingredients for an interesting natural experiment; all that was needed was a network of wires large enough to catch nature in the act of inducing currents. The thirty-thousand-mile-long telegraph network available in 1848 provided just the right technology for the experiment, and during the next few years telegraphists caught much more than simply the dots and dashes they had bargained for. For a long time they had no clue what was going on in their wires.

During the aurora of November 17, 1848, the clicker of the telegraph connecting Florence and Pisa remained stuck as though it had become magnetized, even though the receiving apparatus was not in action at the time. This could only happen if an electric current from some outside source had flowed through the wires to energize the electromagnet. Telegraphers elsewhere began to notice that their lines mysteriously picked up large voltages that caused their equipment to chatter as well, with no signal being sent. Much of this was soon attributed to the long wires picking up lightning discharges in their vicinity, and the solution was simply to erect lighting rods on the telegraph poles. This solution seemed to work for some of the problems but failed to cure all of them. American telegraphists had only a short time to puzzle over atmospheric electricity on their one-thousand-mile lines when, in 1859, the Great Auroras of August 28 and September 4 blazed forth and lit up the skies of nearly every major city on the planet. It was one of the most remarkable displays ever seen in the United States up until that time, and its effects were simply wonderful.

These aurora were so exceptional that the *American Journal of Science and Arts* published no fewer than 158 accounts from around the world describing what the display looked like, the telegraphic disruptions they produced, and assorted theoretical speculations about what was causing them in the first place. Normal business transactions requiring telegraphic exchanges were completely shut down in the major world capitals. In France, telegraphic connections were disrupted as sparks literally flew from the ends of long transmission lines charged to thousands of volts. There were even some near electrocutions. In one instance, Fredrick Royce, a telegraph operator in Washington D.C., reported,

> During the auroral display, I was calling Richmond, and had one hand on the iron plate. Happening to lean towards the sounder, which is against the wall, my forehead grazed a ground wire. Immediately I received a very severe electric shock, which stunned me for an instant. An old man who was sitting facing me, and but a few feet distant, said he saw a spark of fire jump from my forehead to the shoulder.

While a silent battle was being waged between telegraphists and aurora, Alexander Graham Bell, in 1871, uttered the first telephonic sentence in his laboratory, "Mr. Watson, come here. I want you." In less than a year, Watson and Bell completed the first long-distance phone call between Cambridgeport and Boston, using borrowed telegraph lines. Meanwhile, as if to celebrate this event, the Great Aurora of February 4, 1872, colored the skies. Again, reports could be found in the newspapers and science journals of powerful voltages induced upon telegraph lines. During the November 17, 1882, Great Aurora, the telephone lines of the Metropolitan Telephone Company refused to work for most of the business day. Disruptions were also reported on the cables to Cuba and Mexico. The Chicago stock market was severely affected all day, as the business community suddenly discovered their vulnerability.

There was little that anyone could really do about this interference problem. By the time impacts were identified, it was already far too late to rethink deployment of the technology. The famous September 1859 storm lashed the Earth at a time when telegraphy had already become a transcontinental reality, displacing the Pony Express with thirty thousand miles of line strung up on trees and poles. Telephony was born nineteen years later, and its vulnerability was put to the test during the November 18, 1882, solar storm.

By 1901 there were over 855,000 telephones in service in the "Bell Telephone System." It seemed as though the telephone industry had taken the country, and the world, by storm. Everyone wanted their own private line, and the only limiting factor to the spread of this technology was how quickly Bell could cut down trees to make telephone poles and wire city blocks or whole towns into the growing national circuitry. Today, the same public urgency exists in the cellular telephone market. Everyone wants their own cellular phone, and telecommunications companies can't launch satellites or erect relay towers fast enough to keep up with the demand.

No sooner had some considerable money been spent on wiring the world for telegraph and telephone, but a still newer technology appeared in full bloom literally out of nowhere. Guglielmo Marconi in 1895 tinkered together the first spark gap radio wave transmitter and receiver in the garden of his father's estate. The design for it was so simple, Marconi wondered why scientists hadn't developed it themselves in their own laboratories, where all the parts were readily available. Instead of transmitting and receiving electrons flowing in a wire, it was the "wireless" emission of electromagnetic radiation by sparks that carried the messages. The spark intervals could send Morse Code signals as easily as through wire. By 1905, there were over one hundred wireless telegraph transmitters in the United States, with transmission ranges of five hundred miles. There were also some seven million telephones in service on the same wires that once carried telegraphic messages.

So what did people do with this new technology? Many people sure didn't use it very responsibly. Unlike the telephone or telegraph where the ends of the lines are geographically known, for wireless broadcasts, everyone is anonymous unless they choose to identify themselves. As historian Edward Herron wrote in *Miracle of the Air Waves: A History of Radio*,

> [Amateurs] thrilled to calls for help from sinking ships . . . and were not above creating synthetic excitement . . . sending out false messages that caused international distress, confusion, and waste of time and resources. . . . Commercial stations depending on the dollar revenue from the dots and dashes, were constantly at war with the amateurs who rode ruthlessly into the same wavelengths, causing havoc with the commercial messages.

This forerunner to modern computer "hacking" was the main reason why the U.S. government had to step in and put an end to the unruly amateur broadcasts in 1917. Once World War I had concluded in 1919, the embargo was lifted, and the pace of radio technology research exploded like champagne out of a bottle. Almost overnight, the technology for transmitting direct voice messages became a reality. The first commercial radio station, KDKA, owned by Westinghouse opened for business on November 2, 1920, to a hungry crowd of over 30,000 amateur wireless operators who had cobbled together their audio receivers as home hobbyists. Two years later, there were 1.5 million sets in use, and by the end of the decade there were radio sets in 7,500,000 homes. This phenomenon had taken eight years to escalate to this level, whereas telephone service had taken thirty-seven years to reach the same number of homes. Today's stampede of people onto the Internet is only the most recent of many waves of colonization of new high-tech niches that have opened up during this century. Even today, with cell phones and the Internet, we still marvel at the breathtaking speed with which new communication technology becomes commonplace.

Most of the broadcasting during the 1920s was done at long wavelengths, but by 1925 the Navy got involved with shortwave broadcasting because it could be received over long distances with little interference. It could also be transmitted during the daytime, unlike the then popular long-wave transmissions. Wars are fought day and night, so there was tremendous pressure to push transmission technology to higher frequencies and shorter wavelengths where daytime "bounce" was possible. Ironically, the shortwave radio frequencies would drive communication into the very domain that made it a victim of solar interference. Now, whenever aurora dominated the sky, geomagnetic storms were brewing, or the Sun was throwing out flares like electromagnetic thunderbolts, their impacts would appear in many different guises and across the entire spectrum of communications technology. By the time Solar Cycle 17 began, in 1933, twenty-three million homes (70 percent of the total homes) had shortwave radio receivers, and Americans listened to nearly one billion hours each week of broadcasting. Television receivers operating in the newly conquered megacycle radio spectrum were already being field-tested by several manufacturers and were expected to be available to the consumer within a few years. Shortwave interruptions were an increasingly common annoyance during daytime broadcasting, but their origins in distant solar flares were not recognized until 1937.

Meanwhile, as communication technology evolved, the Sun continued to dazzle us with both beautiful aurora and a frustrating barrage of problems. The Great Auroras of January 1938, March 1940, and February 1956 were seen in Europe and as far south as Sicily. British citizens in 1938 were awestruck by the biggest display they had seen in fifty years and actually thought the intense red colors meant that London was aflame. While some viewers watched with dread as the aurora danced throughout Europe, crowds in Vienna awaiting the birth of Princess Juliana's baby cheered the January aurora as a lucky omen. Catholics in the millions around the world were convinced that the January 1938 Great Aurora had been foretold by three young girls in Fatima, Portugal on October 13, 1917, in what many would later call a miraculous visitation by Mary, the mother of Jesus. The "lights in the sky" were to come soon before another major war, which in March 1938 began when the Nazis occupied Austria. The geomagnetic effects that accompanied the spectacular March 1940 aurora caused interruptions in millions of Easter Sunday calls to grandma between 10:00 A.M. and 4:00 P.M. Even the executive curator of the Hayden Planetarium, William Barton, had to go on a nationwide radio hookup to explain what was going on. According to the *New York Times*, the February 1956 Great Aurora included "one of the most intense blasts of cosmic rays ever recorded by scientists" up until that time, causing a spectacular red aurora in Alaska that colored the sky crimson. But while scientists and the public were being entertained aboveground, a far more serious chain of events was unfolding beneath the sea. A full-scale naval alarm had been raised for a British submarine, which was thought to have disappeared. The *Acheron* had been expected to report her position at 5:05 EST while on arctic patrol. When it failed to do so, emergency rescue preparations were begun. Ships and planes began the grim task of searching the deadly, ice cold waters between Iceland and Greenland, but no trace of flotsam or jetsam from the submarine was ever seen. Then, as the auroral activity began to subside, the "missing" submarine turned up four hours later when its transmissions were again picked up.

Military interests in space weather conditions also came into conjunction with one of the most celebrated events in the history of the twentieth century: D day. No expense was spared to make certain the Allied invasion didn't also coincide with a shortwave blackout from an errant solar flare. Walter Orr Roberts, from his solar observatory above Fremont Pass in Colorado, would file daily "Top Secret" reports on the

Sun's current activity. Thousands of miles away, military planners scrutinized these brief missives to make sure the timing of the invasion was "just right" and in a slot, free of any possible solar mischief. His lifelong studies of the Sun, however, eventually led to his death in 1990 from melanoma at age seventy-four. From his mountaintop observatory ultraviolet rays bathed him and planted the hidden seeds that later consumed him, but not before he had founded the National Center for Atmospheric Research in Boulder.

The February 10, 1958, Great Aurora painted the skies over Chicago and Boston in vivid reds and greens, following close behind a terrible snow storm in upstate New York and another Redstone rocket launched by the U.S. Army. In a foretaste of what would become a common, and expensive, problem decades later, the *Explorer 1* satellite launched two weeks before the aurora suddenly lost its primary radio system. The geomagnetic activity also knocked out telecommunications circuits all across Canada, and, although it was not visible in the New York area, it was so brilliant over Europe that it aroused centuries-old fears that some kind of battle or catastrophe was in progress. The Monday storm cut off the United States from radio contact with the rest of the world, following an afternoon of "jumpy connections" that ended with a complete radio blackout by 3:00 P.M., although contact with South America seemed unaffected. By evening the conditions had not improved, and radio messages to Europe could only occasionally be sent and received.

Radio and TV viewers in the Boston area, however, were also having their own amusing problems. For three hours, they fiddled with their TVs and radios as their sets went haywire, at times blanking out entirely or changing stations erratically. Channel 7 viewers began getting channel 7 broadcasts from Manchester, Vermont, while channel 4 viewers received ghostly blends of the local Boston station and one in Providence, Rhode Island. On this Monday evening, families had put their children to bed and were wrapping up their evening's activities with the *Lawrence Welk Show* at 9:30 P.M.. Others were waiting to watch a nationally broadcast TV movie, *Meeting in Paris*, on channel 4, or listen to a boxing match. What they hadn't counted on was that they would get to do both at the same time. Jane Greer played the ex-wife who asks her former spouse, played by Rory Calhoon, to smuggle her new husband out of France. Instead, they discover "that the old spark of romance is still alive when they have a strange encounter in Paris." During a passionate love scene, the audio portion of the movie was replaced by the blow-by-blow details of a boxing match:

> Smith gave him a left to the jaw and a short right hook to the button.
>
> But darling we love each other so much.
>
> A left hook to the jaw flattened Smith and he's down for the count.
>
> Kiss me again, my sweet.

Whenever major aurora took the skies, newspapers seemed to enjoy announcing the many problems they could create, and, before 1960, they usually put the news on the front page. Television was still very new, and most people had little idea what to expect from it in terms of clarity or reliability. Today, as the news media suffer from information overload, we seldom hear of telephone or shortwave interruptions making any impact on us, at least not the way they used to. But that doesn't mean the underlying problems have gone away. Aurora are in most instances only a poor indicator of other invisible events taking place within the magnetic region surrounding the Earth–the magnetosphere. Some of these disturbances can even affect our technology without producing an aurora at all. In nearly all such geomagnetic storms, some aspect of our technology is affected. In fact, "solar storms" are now seen as a problem for many of the key systems in our technological infrastructure. One of these is as close to you as the light switch on your wall.

Part II

The Present

4 Between a Rock and a Hard Place

I am a lineman for the county
And I drive the main road
Searching the Sun for another overload

—Jimmy Webb, "Wichita Lineman," 1969

Along the Pacific Coast, from Oregon to Baja California, it was turning out to be another sweltering day. Normally, the more moderate temperatures in the Northwest allowed extra power to be available to feed millions of air conditioners in the south, but in August 1996 this would not be the case. Temperatures climbed into the triple digits as the Sun rose higher in the sky. Already hot power lines from the Oregon power grid began to overheat as they carried much of the 21,450 megawatts needed to support the thousands of air conditioners that came on-line every minute. On August 10, 1996, an overheated 500,000-volt power line between Keeler and Allison sagged an inch too far. A powerful and blinding arc of electricity jumped into the branches of a nearby hazelnut tree. Automatic sensors in the huge Pacific intertie sensed a problem and began shutting the system down. In an instant, six million people found themselves without power for up to several days.

In the San Francisco Bay Area, the outage started at 4:00 P.M. and lasted over six hours. It was the second summer blackout for California residents in less than two months since an earlier July incident that affected fourteen states. This time, only six states were involved, including Texas and Idaho. BART subway lines were without power and most of the cable cars in San Francisco could not be budged. Six thousand passengers were stuck in planes that were taxing to their gates at San Francisco International Airport and airports in Oakland and San Jose. Huge clouds of black smoke belched from the Chevron petroleum refinery in Richmond as equipment malfunctioned. Some shoppers at supermarkets in San Francisco actually enjoyed walking down darkened isles and thought the experience "surreal and dreamlike." In the southern end of the state, a scenic ten-mile stretch of beach was flooded by raw sewage as the Hyperion Treatment Plant poured six million gallons of untreated effluent into the ocean. The Republican National Convention was nearly routed when the lights blinked but stayed on. Delegates, however, returned to darkened hotels and had to navigate massive traffic congestion to get there after a long day. Merchants throughout the affected six-state region were forced to calculate merchandise taxes by hand for the first time in a decade. Las Vegas casinos found themselves plunged into darkness, without air conditioning or working slot machines.

You would be surprised how often blackouts of one kind or another manage to rumble through our country. In some regions, they are a yearly summer ritual affecting tens of thousands of people for hours at a time. In other regions, momentary fluctuations in voltages, called "sags," cause blackouts lasting only a few minutes. You don't even notice they have happened unless you arrive home and wonder why all the electric clocks are now blinking "12:00." Major blackouts involving millions of people have been mercifully rare. The last major power outage to rock the United States happened on November 9, 1965, and led to a presidential investigation of the electric power industry.

A variety of temporary outages during the May–June 1998 Midwest heat wave cost steel manufacturers tens of thousands of tons of steel production and millions of dollars in lost profits. Companies can purchase emergency power, but local electrical utilities charge them rates that are hundreds of times the regular rates. Even minor fluctuations in electric voltage, which happen on a daily basis in many regions, can stop newspaper presses cold and cost a printing company tens of thousands of dollars a year in extra labor and lost paper.

In virtually all of these cases, the cause of the blackout is something rather easy to visualize. A particular component might have failed, or a power line might have been downed, leading to a cascade of breakdowns that swept through a utility system in literally a few seconds. It's much harder to imagine the same aggravating problem happening because of a distant solar storm. It is such a counterintuitive idea that, even when you are in the middle of such an event, the Sun is the last thing you think about as the cause of the problem. It's much easier to point a finger at some dramatic natural phenomenon like a lightning strike, a downed tree ripping down a power line, or even human error. But like so many other freak events we hear about these days, eventually even rare cards get dealt once in a while.

The first public mention that electrical power systems could be disrupted by solar storms appeared in the *New York Times*, November 2, 1903, "Electric Phenomena in Parts of Europe." The article described the by now usual details of how communication channels in France were badly affected by the magnetic storm, but it then mentions that, in Geneva Switzerland, "all the electrical streetcars were brought to a sudden standstill, and the unexpected cessation of the electrical current caused consternation at the generating works where all efforts to discover the cause were fruitless."

Of course they were fruitless. By the time the investigation began, the celestial agent responsible for the mess had already left town. In a repeat story a decade later, we hear about another aurora seen in Scandinavia on January 26, 1926: "A breakdown of electrical power and light caused considerable inconvenience in Liverpool yesterday. Mr. Justice Swift was trying a burglary case when the lights failed, and the hearing proceeded without lights."

The United States and Canada are geographically closer to the north magnetic pole, and to latitudes where auroras are common, than most areas in Europe including Scandinavia, so we have a ringside seat to many of these displays whether urban dwellers can see them or not. This also makes us especially vulnerable to geomagnetic disturbances and their auroral co-conspirators, and we experience these far more often than our European counterparts. Electrical power companies have supplied a widening net of consumers since the first 225-home lighting system was installed in 1882 by Thomas Edison. The stealthy effects of geomagnetic disturbances took a very long time to reach a threshold where their impact could actually be registered. A few extra amperes from celestial sources went entirely unnoticed for a great many years. The watershed event came with the March 24, 1940, solar storm, which caused a spectacular disruption of electrical service in New England, New York, Pennsylvania, Minnesota, Quebec, and Ontario. By then, it was entirely too late to do much about the problem. With a little detective work, you can uncover many other mentions of solar storm-related electrical problems in New England, New York, Minnesota, Quebec, and Ontario. Almost like clockwork, whenever the sunspot cycle is near its peak and a major Great Aurora is spotted, we get a wake-up call that our electrical power system is not as secure as we would hope it might be. We have already heard about the Quebec blackout of 1989, but there have been many other electrical problems before then and afterward.

The Great Aurora of August 2, 1972, triggered surges of 60 volts on AT&T's coaxial telephone cable between Chicago and Nebraska. Meanwhile, the Bureau of Reclamation power station in Watertown, South Dakota experienced 25,000-volt swings in its power lines. Similar disruptions were reported by Wisconsin Power and Light, Madison Gas and Electric, and Wisconsin Public Service Corporation. The calamity from this one storm didn't end in Wisconsin. In Newfoundland, induced ground currents activated protective relays at the Bowater Power Company. A 230,000-volt transformer at the British Columbia Hydro and Power Authority actually exploded. The Manitoba Hydro Company recorded 120-megawatt power drops in a matter of a few minutes in the power it was supplying to Minnesota.

Despite the aggravation that many people had to endure during the 1972 Great Aurora, it was actually a very good year for electrical power in North America. We had far more available power than we used, even during peak load conditions in the summer. Air conditioners were still rare in the urban world. With each passing year, however, we have found more uses for electricity than the pace with which we have created new supplies for it. The advent of the personal computer alone has added more than 3,000 megawatts per year to domestic power consumption since the 1980s. Steadily, the buffer between electrical supply and consumer demand has been whittled away. Solar and geomagnetic storms continue to happen, but now there is much less wiggle room for power utilities to find, and purchase, additional power to tide them over during outages or sags. We don't build new power plants with the fervor we used to during the go-go sixties. No one wants them in their community, and those ugly power towers 100 feet tall are an eyesore to our suburban sense of aesthetics and their way of impacting on our property values. So now utilities have learned how to buy and sell dwindling reserves of available power across states and whole regions within a stagnating production climate.

As North America has evolved into a unified power-sharing network of regions, each buying and selling a diminishing asset, U.S. domestic power has become more vulnerable to solar storms buffeting the power grid in the more fragile northern-tier states and Canada. So long as one region continues to have a surplus at a time when another region needs a hundred megawatts, power is "wheeled" through one-thousand-mile power lines to keep supply and demand balanced across the grid. In 1972, a typical utility might need to conduct only a few of these electromagnetic transactions each week. Now, it is common for thousands to be carried out, often by computer, in much the same way that stocks are automatically traded on Wall Street. Solar storm disturbances that once hid under the cloak of an adequate power margin are now exposed like the ribs of some malnourished behemoth. Only a strong "kick" by the Sun is needed to cause an avalanche of outages that can affect not just individual towns but entire geographic regions.

With communications technology, it is not too hard to figure out how aurora and magnetic storms do their damage. With electrical utilities, however, there are several things going on at once that make it hard to follow how an electrical system has to fail in order to cause a blackout. Unlike our understanding of how telegraph and telephone lines are affected, what we know about power lines and solar storm impacts is much more recent. Although there have been reports of major power surges from auroral currents since the Great Aurora of 1940, routine measurements of induced power grid currents in places like the United States and Scandinavia were not begun until at least the 1970s. The results were quite surprising. Michael Faraday showed in the 1800s that changing magnetic fields can induce electrical currents in a wire. What these power grid measurements showed was that geomagnetically induced currents (what engineers call GICs) in the ground could cause hundreds of extra amperes to flow in some lines and induce voltages as high as 100 volts per mile (for example, during the March 1940 storm). Considering that we now have some 180,000 miles of high-voltage power lines in North America, this is quite a large collector for even the smallest squall that wafts by on the solar wind.

All you have to do is take a Sunday drive in the country, and you will see tall towers marching like giants from horizon to horizon. Looped between them are cables as thick as your arm, carrying hundreds of thousands of volts of electricity. You would think that geomagnetic currents would not be much of a problem. They could even be considered a new source of free electrical energy since carrying electricity is what power lines are designed to do. The problem is that the geomagnetic currents are the wrong types. They are similar to the kind of electricity you get in a battery called direct current (DC). But, the electricity you get from a plug on the wall is alternating current (AC). In a system designed to carry AC electricity, DC currents are very bad news.

The electrical power grid is composed of many elements, and you can think of it as a set of rivers flowing overhead. Large rivers carry the electricity from distant generation stations (dams, hydroelectric facilities, and nuclear plants) on supply lines of 138,000 volts or higher. These are carried as three cables suspended atop 100-foot tall towers that you will see out in many rural areas. These supply cables terminate at regional substations where the high voltages are converted into lower voltages from 69,000 volts to 13,800 volts. These lines then enter your neighborhoods atop your local telephone poles where a neighborhood transformer steps this voltage down to 220 and supplies a dozen or so individual houses. Like an orchestra, this entire network acts as a single circuit that has complex ways of vibrating electrically, depending on the kinds of loads it is serving at a given moment. Typically, in North America, various components such as transformers, capacitors, and other devices can split the 60-cycle oscillations into harmonics at 120, 180, and even higher vibrations, suppressing some of them or amplifying others.

But how do GICs affect house-sized transformers in the first place? It seems absurd that a few dozen extra amperes of electricity can make any difference to a transformer delivering thousands of volts of electricity. For a transformer to operate normally, the current and voltage oscillates in a specific phase relationship that has to do with the iron-steel content of the core and the geometry of the transformer. Like two sets of ocean waves lapping up onto the beach 60 times a second, the voltage and current waves traveling down a line can be out of synch with each other in AC electricity. Depending on the kinds of loads the line is supplying, from electrical motors to heating elements and fans, the voltage and current can get pulled out of synch to greater or lesser degrees.

When GICs enter a transformer, the added DC current causes the relationship between the AC voltage and current to change. Because of the way that GIC currents affect the transformer, it only takes a hundred amperes of GIC current or less to cause a transformer to overload, or "saturate," during one-half of its 60-cycle operation. As the transformer switches 120 times a second between being saturated and unsaturated, the normal hum of a transformer becomes a raucous, crackling whine. Regions of opposed magnetism as big as your fist in the core steel plates crash about and vibrate the 100-ton transformer nearly as big as a house, in a process that engineers call magnetostriction.

Magnetostriction is bad news for a transformer because it causes both mechanical and thermal damage. It generates hot spots inside the transformer where temperatures can increase very rapidly to hundreds of degrees in only a few minutes. Temperature spikes like these can persist for the duration of the magnetic storm, which itself can last hours at a time. During the March 1989 storm, a transformer at a nuclear plant in New Jersey was damaged beyond repair as its insulation gave way after years of cumulative GIC damage. Allegheny Power happened to be monitoring a transformer that they knew to be flaky. When the next geomagnetic storm hit in 1992, they saw the transformer reply in minutes and send temperatures in part of its tank to more than 340 F (171 C). Other transformers have spiked fevers as high as 750 F (400 C). Insulation damage is a cumulative process over the course of many GICs, and it is easy to see how cumulative solar storm and geomagnetic effects were overlooked in the past. You only see them if you open up a transformer and inspect it directly, or keep careful track of the amount of gas that has accumulated in the coolant liquid.

Outright transformer failures are much more frequent in geographic regions where GICs are common. The northeastern United States, with the highest rate of detected geomagnetic activity, led the pack with 60 percent more failures. Not only that, but the average working lifetimes of transformers was also found to be shorter in regions with greater geomagnetic storm activity. The rise and fall of these transformer failures even follows a solar activity pattern of roughly eleven years.

The problem doesn't end with something as dramatic as a transformer heating up and failing catastrophically. Even nondestructive GICs also affect the efficiency with which a power grid is transmitting power. Because we have less power available to support the new demands placed on the power grid, engineers must constantly monitor the efficiency at which power is being generated and delivered. A 1-percent drop in efficiency can mean megawatts of power wasted and millions of dollars in revenue lost.

It isn't just the transformers and lines that can make you susceptible to GICs; the very ground under your feet can act as an invisible co-

conspirator. But rocks have their own patterns of resistance. Igneous rocks, which are common in most areas of North America, are particularly troublesome. If your power plant is located over an igneous rock stratum with high resistance (low conductivity), any geomagnetic disturbance will cause a significant change in the voltages of your local ground. The bigger this change in ground voltage, the stronger will be the GIC currents that flow into your transformers. This is how geomagnetic storms enter the power grid. Typical daily GICs can cause about 5–10 amperes of current to flow into the ground connection of a three-phase transformer, but severe geomagnetic storms can cause 100–200 amperes to flow.

A conservative estimate of the damage done by GICs to transformers by Minnesota Power and Light was $100 million during a solar maximum period. This includes the replacement of damaged transformers and the impact of shortened operating lifetimes due to GIC activity. This doesn't sound like much, especially compared to the outfall from other natural calamities such as hurricanes and tornadoes, but the implied level of electric service disruptions for the dozens of transformers taken out of service is considerable. It is at this point that the effects of invisible geomagnetic storms are greatly multiplied out of proportion to the seemingly innocuous currents they induce in a house-sized transformer. John Kappenman, an electrical engineer at Metatech Corporation, reflects on all of this by noting that

> the evolving growth of the North American transmission grid over the past few decades has made the grid, along with the geographical formations occurring in much of North America, the equivalent of a large efficient antenna that is electromagnetically coupled to the disturbance signals produced by [GICs]. . . . Yet monitoring is only being done at a handful of the many thousands of possible GIC entry points on the network.

Large transformers cost $10 million, and can require a year or more to replace if spares are not available. During a transformer failure, an affected utility company will have to purchase replacement power from other utilities for as much as $400,000 per day or more, quickly wiping out the profits of many electrical utilities. Oak Ridge National Laboratories, meanwhile, estimated that a solar storm event only slightly stronger than the one that caused the Quebec blackout in 1989 would have involved the northeastern United States in a cascading blackout. The experts figured that about $6 billion in damages and lost wages would have resulted from such a widespread involvement. The North American Electric Reliability Council (NERC) placed the March 1989 storm event in a category equivalent to Hurricane Hugo or the 1989 Loma Prieta Earthquake in San Francisco. But many consultants for the power industry dispute NERC's estimate, saying that it may actually be much too low. The $6 billion may not properly include collateral impacts such as lost productivity, spoiled food, and a myriad of other human costs that could easily run the losses into the tens of billions of dollars.

The average person has never experienced a brownout or blackout caused by a solar storm event, and this is in large measure due to the intrinsic robustness of the power grid technology. It's also a matter of old-fashioned good luck! We are entering a new era of substantial increases in electrical power demand that drives power grids to work near their

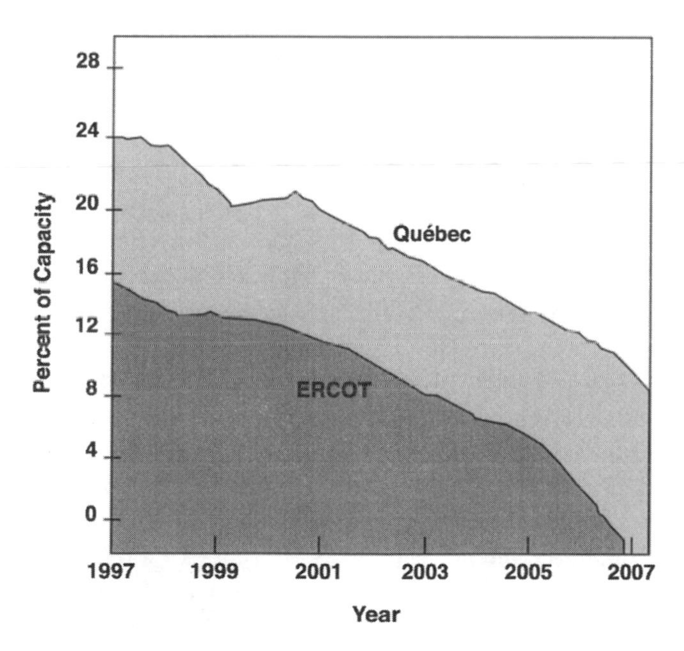

FIGURE 4.1 Electrical capacity margins for two major regions of North America showing the sharply declining balance between available supplies and demands through 2007.

SOURCE: *North American Electric Reliability Council*

maximum capacities with dramatically eroded operating margins and for many more days during each year. Large purchases of power from the more vulnerable Canadian power grid places domestic electrical reliability in increasing jeopardy.

Because of the increased electrical demand over the last decade, and the insufficiency of communities to build new power plants, North American citizens have created a potentially unstable balance between supply and demand. In 1998, for example, the peak power demand was 648,694 megawatts, compared to the 737,855 megawatts that was available in actual power plant capacity, a margin of only 16 percent. A decade earlier, this margin was substantially higher. The NERC has found that we are rapidly reaching a critical condition: the amount of available electrical capacity in excess of peak demand will shrink from 19 percent during peak summer load conditions in 1995 to 10 percent by 2004. Over the same period of time, the margins for specific regional power grids such as the ERCOT Interconnection of Texas will shrink from 20 percent to a few percent if none of the proposed power plants are built and if the expected power plant retirements occur. This means that there is less electricity available for utilities to buy during geomagnetic storms when the power grid is working under peak demand conditions. During the March 1989 Quebec blackout, Hydro-Quebec could purchase, for few days, thousands of megawatts of "excess" power from other states. As we enter the twenty-first century, a similar blackout at the same time of year may take much longer to resolve because less power will be available to purchase. That means that, on average, more people will be in the dark, and bundled up against the cold or sweltering in the heat, for a longer period of time. As the NERC noted in a summary of its report, "Lower capacity margins can diminish the ability of the bulk electric supply systems in North America to respond to higher-than-projected customer demands caused by extreme weather and unexpected equipment shutdowns or outages." Energy Secretary Bill Richardson does not find the U.S. power system in the best of conditions, especially after deregulation, which began in April 2000. In a *Washington Post* article, he comments, "The state of the American power system is bleak. We have a booming economy with an antiquated power system. We have inadequate generating capacity, inadequate transmission capacity and there are cutbacks on energy efficiency."

In the eastern United States, only 24,400 megawatts of new generating capacity will come on-line by 2002, but by then the projected demand will have grown an additional 36,000 megawatts, and perhaps as much as 47,000. As we are forced to operate our electrical utilities with diminishing margins for emergencies, we become much more vulnerable to any kind of outage of equipment, no matter what the cause or how seemingly infrequent. Geomagnetic storms can then grow to become the proverbial straw that breaks the camel's back.

The United States has only recently warmed up to GICs as a significant problem requiring serious attention. Countries such as England, Scotland, and Finland have been aggressively working on GIC mitigation since 1982. In England, for example, they have a single power utility that includes Wales and also connects with France. During the 1980s, they endured a number of strikes by coal miners that triggered electrical supply problems and sensitized the populace to just how vulnerable their lifestyles are, even to intermittent losses of power. When British electrical engineers and scientists brought GICs to the table, utility managers were much more interested in mitigating even these rare impacts. Having been beat around the head and shoulders by the public, and by politicians, for outages they could not control, the British power industry welcomed any new insight that might keep their customers happy.

In 1991, Bill Feero, an electrical engineer from the Research and Management Corporation in State College, Pennsylvania, developed a real-time monitoring system called Sunburst, which could measure the GIC currents at hundreds of locations across North America and Europe. All that participating electrical utility companies such as the Potomac Electric and Power Company, Virginia Electric and Power Company, and Baltimore Gas and Electric needed to do was to install a passive measuring device on selected transformers at their substations. These devices, no bigger than a bagel, transmit by phone line minute-by-minute GIC current measurements to Sunburst headquarters in Pittsburgh. In essence, the system turns the global power grid into a vast space weather gauge. When the readings exceed preset levels, warnings can be sent to the participating power companies to alert them to conditions that could lead to an equipment outage. Moreover, this equipment has also made several important discoveries of its own.

Before the advent of Sunburst, many engineers thought GICs could cause power transformer failure under only extreme conditions and, generally, involving only the primary "60-cycle" electrical responses of the equipment. Now it is recognized that the higher harmonics of this 60-cycle frequency can also do significant damage by causing stray currents to flow in large turbine generators. Also, capacitor banks that help maintain network voltages can be tripped and taken off-line by the voltage spikes produced by these harmonic currents.

One problem with real-time power system monitoring is that, although it is far better than being caught unawares, once a GIC starts to happen, you have precious few seconds to do anything meaningful except perhaps go outside for a smoke. Severe storms like the one that caused the Quebec blackout are preceded by very normal conditions, and within a few seconds the GICs rise sharply to their full levels of hundreds of amperes. Local real-time measurements alone will probably not be enough by themselves in guiding plant managers to take meaningful action, although the information can be used in a postmortem or forensic mode to let plant managers know which devices are the most vulnerable. Another approach is to try to forecast when GICs will happen. This is not as impossible as it seems.

John Kappenman takes satellite data from the Advanced Composition Explorer (ACE) a NASA satellite located one million miles out in space and feeds it into a sophisticated computer program called PowerCast. Within seconds, a complete picture appears of the expected GIC currents at a specific transformer a half-hour later. An electrical utility company running PowerCast can look at any line, transformer, or other component in their system and immediately read out just what it will do when the solar wind hits the Earth traveling at a million miles per hour. With thirty minutes to spare, it is now possible to put into action a variety of countermeasures to gird the grid from failure.

To make the forecast, satellite data tells a program what the direction of the solar wind magnetic field is at a particular instant. If this polarity is opposite to that in the Northern Hemisphere, a geomagnetic event will be spawned. This event will cause an electrical current, called an "auroral electrojet," to flow in the ionosphere. As this current flows overhead, it causes a sympathetic current to flow in the Earth. PowerCast calculates, from the satellite data, the expected strength of the electrojet current, then the amount of induced ground current based on a detailed model of the rock conductivity under a particular transformer or power line. This is all done on a PC computer in real time.

The lynchpin in this powerful system of GIC forecasting is the *ACE* satellite and its onboard solar wind monitor. At a distance of one million miles toward the Sun, its instruments report on the minute-by-minute changes in the density, speed, and magnetic orientation of the solar wind.

For decades space scientists have known that when the magnetic polarity of the wind dips southward it triggers violent instabilities in the Earth's magnetic field in the Northern Hemisphere. When like-polarity conditions prevail, the magnetosphere receives a constant but firm pressure from the wind in much the way that two magnets with the same poles facing each other push apart. But when the polarities are opposed, fields intermingle and reconnect into new shapes in a dynamic process. Currents flow in the polar regions of the Earth, and it is these currents that cause GICs to bloom in transformer cores like dandelions on the ground.

As scientists and engineers have grown more familiar with what to look for when solar storms are brewing, it has also become easier to reveal other ways in which these storms can invade our technology and power systems. There are also many ways that their impacts can be hidden or camouflaged by unrelated events.

5 "We're Not in Kansas Anymore!"

> No master mariner dares to use it least he should be suspected of being a magician; nor would the sailors venture to go to sea under the command of a man using an instrument which so much appeared to be under the influence of the powers below.
>
> —Guiot of Provins, ca. A.D. 1205

The *Exxon Valdez* left harbor on March 23, 1989, and within hours unleashed an ecological catastrophe as it ran aground on Bligh Reef 12:04 A.M. on March 24. The spilling of eleven million gallons of oil triggered a $5.3 billion lawsuit in a highly publicized court case. A decade later, the damage to the Prince Williams Sound is still evident if you literally scratch the surface of the ecosystem. The investigation focused on the circumstances leading up to the grounding, the absence of the captain from the bridge, and the failure of the third mate to follow a proper course, but there may have been other factors at work as well.

The powerful geomagnetic storm that triggered the Ontario blackout on March 13 was not the only event that rocked the magnetosphere that month. Ten days later, a secondary storm began fourteen hours before the *Exxon Valdez* ran aground. The *IMP-8* research satellite recorded a powerful surge of high-energy electrons and protons lasting twenty-four hours. Meanwhile, images from the *Dynamics Explorer* satellite showed a bright crown of aurora girding the north polar zone, especially in the nighttime sector that included Canada and Alaska. Had the skies been clear over Valdez Harbor on March 24, sky watchers would have seen a marvelous Northern Lights display.

The displays, by themselves, have little consequence for navigation, but they do signal that powerful currents of electrons are flowing in the upper atmosphere, and these electrojet currents make themselves felt, magnetically, on the ground. Magnetic field data from middle-latitude observatories traced significant changes in the vertical component of the geomagnetic field. The "Dst" storm-time geomagnetic index, which tracks these changes, like a barometric reading, pitched up and down between March 22–25 as ionospheric currents flowed strongly but erratically and reduced the Earth's field by up to 1 percent vertically. Meanwhile, magnetic observatories at Sitka, Barrow, and College, Alaska recorded up to one-degree changes in the direction of magnetic north. The *Exxon Valdez* ran aground in the middle of one of these magnetic excursions that occurred between 11:00 P.M. on March 23 and 12:30 A.M. on March 24.

After leaving the harbor, the *Exxon Valdez*, like other ships ahead of it, had to navigate through a narrowing of the channel. The Columbia Glacier ice flows had calved many icebergs, and these house-sized mountains of frozen water were now flowing into the channel, constricting it to a narrow 1,500-yard wide passage near Bligh Island. Records show that the *Exxon Valdez* made a bearing change at 11:39 P.M. on March 23 which was to be followed on autopilot for four miles before coming around the edge of the Columbia Glacier ice flow. Although there was no magnetic observatory at Valdez, and the nearest ones were at Sitka and College located hundreds of miles away, the magnetic conditions sensed by the three observatories were about the same. A magnetic compass on the *Exxon Valdez* would have detected up to 0.3-degree deviations from the intended bearing, which over a four-mile run would have added up to one hundred feet or so of deviation near the end of the run at Bligh Reef. This is only about one-ninth the length of the tanker. A second course change was planned at the end of the four-mile leg, and it is this one that, investigators demonstrated, came seven minutes too late. Because of the delay, the tanker overshot the narrow slot between the ice flow and the reef, and the rest is history.

Did the magnetic storm cause the grounding of the *Exxon Valdez*? Probably not. Two other ships, the *Brooklyn* and the *Arco Juneau*, had successfully navigated this channel between 10:47 A.M. and 7:22 P.M. on March 23 with no difficulty, and an even larger magnetic storm was in progress between 5:50 P.M. and 6:30 P.M. on that day, with deflections up to one degree. More important, quite a bit has happened to navigation technology in the last fifty years. The Exxon Valdez, one of the most sophisticated ships of its kind at the time, relied on LORAN-C signals to determine its course and not magnetic compass bearings.

This incident, then, seems to be a spectacular case in which a solar storm that presented large magnetic swings was in progress and a ship navigating a narrow channel was grounded, with the two events apparently unrelated. The initial expectation of cause and effect was based on the very logical premise that geomagnetic storms cause significant deviations in compass bearings. However, if no magnetic compass is used, then there can be no navigation impact, even by a major magnetic storm. But we now have to ask ourselves one further question. If the *Exxon Valdez* was immune to direct magnetic disruptions, could the LORAN-C system have been affected instead?

For this to be a viable possibility the storm-time conditions would have to produce significant propagation delays in the shortwave LORAN-C navigation beacons that were used to figure bearings between 11:30 P.M. and 12:30 A.M. Reports on the events leading up to the grounding indicate there were frequent electromagnetic interruptions at the time that the *Exxon Valdez* left harbor. The "blip" representing the tanker on the radar screens apparently faded in and out of detectability as the ship passed Rocky Point, located just outside the Valdez Narrows harbor entrance. This radar problem was well known to the operators at the Valdez Coast Guard Station and, of itself, not a condition that had to do with solar storms. It was simply the consequence of not having a second radar station located near the mouth of the harbor to extend the range of the main installation at Valdez. Moreover, the third mate tested the navigation equipment at 7:24 P.M., before leaving the harbor, and this check apparently included the ship's radar, gyrocompass, automatic pilot, and course indicator. Nothing out of the ordinary was discovered. One would expect that a storm-time process capable of significantly affecting navigation would have showed up just three hours before the grounding and at a time when the geomagnetic disturbances were even stronger.

So we come up empty handed. Even though a cursory examination of the *Exxon Valdez* grounding seemed to turn up an exciting new smoking gun for a very spectacular shipwreck, there appears to be no cause-and-effect link between the key events. Had the *Exxon Valdez*'s navigation been affected by even a one-degree error in its course, the several

100-foot error this would have caused near Bligh Reef would have initially made a difference but wouldn't have avoided the inevitable impact. With a ship traveling at 20 feet per second, it would have just delayed the inevitable by a few more seconds.

Despite the occasional geomagnetic storm, we are fortunate to be living on a planet that has a well-defined magnetic field and has served us as a navigation reference for millennia. Had the situation been otherwise, ancient mariners may have had to steer their daytime courses using only nighttime stars and the daytime Sun as a guide. No one really knows exactly when the first person came up with the idea of using a rock to tell direction. It's hard to imagine the trial-and-error process that could have led up to this discovery. But the history books are pretty clear that many thousands of years ago some nameless soul discovered that a particular kind of rock we now call lodestone (magnetite) does the trick.

The story seems to begin in ancient China, when Emperor Hoang-ti's troops were in hot pursuit of Prince Tcheyeou in 2637 B.C. for reasons that are now lost to us. Ancient Chinese politics was a complex and ever changing arena. The troops eventually lost their way in a heavy fog, so the emperor constructed a chariot upon which stood a figure that always pointed south no matter how the chariot was directed. Nearly two millennia later, the Phoenician sage Sanconiathon wrote, "It was the God, Ouranos, who devised Betulae, contriving stones that moved as having life," and even Homer, about 900 B.C., got into the act by mentioning this remarkable technology in the Odyssey,

> In wondrous ships instinct with mind
> No helm secures their course, no pilot guides
> Like man intelligent, they plough the sea
> Though clouds and darkness veil th' encumbered sky
> Fearless thro' darkness and thru' clouds they fly

During the last thousand years, the "secret weapon" of the Vikings evolved into the familiar magnetic compass that all scouts and ocean navigators rely on to see them to safe harbor. We don't need lodestone anymore. A simple needle balanced midway between its ends suffices to point in a fixed direction. By 1600, William Gilbert, who was the personal physician to Queen Elizabeth, even wrote a book about how the Earth is one giant magnet with distinct north and south poles.

We are living at a time in the history of the Earth when the magnetic north-south field is very nearly aligned with the axis about which the Earth spins each day. We don't know exactly how the Earth creates this field. Geophysicists think that the geomagnetic field is generated near the hot, electrically active core of the Earth where hundred-mile-wide currents of molten nickel flow along the equator. Like many rivers of water on the surface of the Earth, these subterranean currents are not steady either in space or time. Over thousands of years—even near the core of the Earth—things tend to slosh about a bit. If you were standing at the magnetic north pole, you would soon discover that it moves a hundred yards a day, and this forces compass navigators to buy new maps every ten years or so. Map makers and sellers since the eighteenth century enjoy this aspect of geophysics quite a bit and, over time, actually turn a profit from it. There are other, less predictable changes that occur with magnetic bearings if you have the patience to look for them.

In the early nineteenth century, Baron Alexander von Humbolt was one of those intrepid and world-renowned explorers who outfitted expeditions to Africa and elsewhere to catalog rare plants and animals. His popular stature was a combination of the measured studiousness of astronomer Carl Sagan and the down-and-dirty enthusiasm of *Titanic* discoverer Jim Ballard. In fact, the *London Times* regularly published Humbolt's weekly letters from distant lands and jungles detailing his ongoing exploits. On one of his years off from studying wild and exotic fauna and flora, his interests turned to earlier eighteenth-century reports that compass needles didn't always point in the same direction from moment to moment. He and an assistant decided to look into this behavior a bit more.

With a microscope, they made around-the-clock measurements of a compass needle's direction every half-hour for over a year. What they uncovered were the usual, and sudden, erratic swings produced by lightning storms, but every once and a while other mysterious disturbances set their needle gyrating. It didn't take long for them to realize that the strongest of these "magnetic storms" always seemed to happen when the Northern Lights could be seen dancing outside their window or were reported in neighboring lands to the north. This behavior was taken very seriously at the time, because in terms of our monetary system today, billions of dollars of commerce were at the mercy of ships steered by magnetic compass. Within a few years, Humbolt had dozens of "magnetic observatories" across the globe hard at work measuring compass needle gyrations and magnetic storms.

Magnetic storms are not something to trifle with. If you are a navigator, they can cause compass-bearing errors as large as several degrees, so that for up to a full day your bearings are completely unreliable and you might not even realize it. This is especially challenging and risky if you are trying to get through a tight channel in the dark or in inclement weather. The most dramatic impact of geomagnetic storms would be a shipwreck or a plane crashing into a mountainside. Few recorded instances of such tragic events are known; however, there are stories about a ship that ran aground on Bear Island just before World War II, and airplane pilots in Alaska have claimed that some crashes were caused by just such geomagnetic storms. The problem is that historical accounts of geomagnetically induced navigation problems are almost entirely anecdotal. The earliest account is reported in the *American Journal of Science and Arts* by a contributor named, simply, "A. de la Riva,"

> M. de Tessan cites an observation made in 1818 by M. Baral, another French naval officer, on the same coasts of New Holland, who found that he had been making a wrong course from following his needle. . . . But on the evening of the same day, there was a brilliant aurora, and to this he attributes the deviation.

In addition to the many exciting changes in communication technology during the last one hundred years, even the magnetic compass was eventually eclipsed. Navigation could now be provided by a series of strategically placed transmitters that ships and planes could lock onto. Within a few years, the Long Range Aid to Navigation (LORAN) system had all but replaced navigation by stars and compass, although these might be used occasionally as backup aids. LORAN coverage was extended over combat areas and along Pacific supply lines, so that by the end of World War II about 30 percent of the Earth's surface had been covered by LORAN. The system appeared to be resistant to geomagnetic storms, but they did have their weakness.

LORAN shortwave signals could be affected by severe static, and to get a reliable and accurate bearing, you need to measure the arrival times of the signals from three stations. Because these signals have to bounce off the ionosphere to reach you, any changes in the ionosphere cause erratic increases or decreases in the signal's travel time to your ship. This causes course errors just as surely as if you had been using a magnetic compass. For instance, during the March 13, 1989, solar storm activity and even several weeks prior to its onset, the ionosphere was severely affected by this major storm. A number of reports describe how this caused navigation signals to become unreliable for several days. Shortwave radio could not even be used to alert ships at sea to the problems with LORAN. Our previous question about the *Exxon Valdez* and its LORAN-C system now acquires a new and disturbing answer. Its instruments may, indeed, have been affected by the ionospheric disturbances that were probably in progress at the time. We will never know just how much of a factor this could have had in the catastrophe that followed.

Since the early 1990s, a new navigation technology has swept the scene: satellites. The Department of Defense launched twenty-four satellites to make up their Global Positioning System (GPS). Now, with a handset no bigger than a cellular telephone, you can find your instantaneous longitude and latitude no matter where you are on Earth or in orbit. Between five and eight of the satellites are above your horizon at any time, and your handset receives their timing signals. A computer chip inside uses the timing information to triangulate your position to within fifty feet or less. Even so, slight changes in the ionosphere caused by solar storms add minuscule delays to these signals and cause position estimates to vary by hundreds of yards. A major factor that is known to cause ionospheric changes is solar flares.

Solar flares can happen even when aurora or geomagnetic storms are not in progress, and they happen in broad daylight. When a flare erupts on the surface of the Sun facing the Earth, less than nine minutes later a powerful burst of X-ray and gamma ray radiation arrives at the Earth, followed an hour or so later by high-speed energetic particles. This "one-two punch" of matter and energy plays havoc with the daytime ionosphere and causes shortwave dropouts and radio navigation problems that can last for hours. Solar flares and geomagnetic storms are common enough, and navigation difficulties frequent enough, that sometimes the two are conjoined in time to paint a provocative picture.

During the March storm in 1989, the *New York Times*, and many other newspapers, reported a military helicopter crash near Tucson, Arizona killing fifteen people on March 12. It was a moonless night and the pilot was using night vision goggles to navigate their helicopter. The Air Force had flown many missions in this way and it was never cited as a contributing factor in any previous crash. Could this have been a navigation problem from geomagnetic disturbances that caused the Quebec blackout a day later? Investigations of the crash turned up the entirely plausible conclusion that there may have been too little ambient light for the goggles to work properly. The pilot was literally flying blind. Two trains collided head-on in Alberta and microwave-controlled traffic signals may have been affected by ionospheric disturbances. Although space weather expert Gordon Rosloker was prepared to give this testimony, the case was terminated before any aspect of this probable cause could be entered into the court records. In another incident, according to Joe Allen at NOAA, a commercial fishing ship was seized by the Australian Coast Guard in forbidden waters. The captain was arrested and ordered to stand trial. Navigation errors from the March 1989 geomagnetic storm may also have had a hand in this incident as well.

On March 11, 1989, at 1:10 P.M., Air Canada Flight 1363 crashed soon after takeoff, killing sixty-nine people during a snowstorm with one-half-mile visibility. Could *this* have been caused by navigation problems? Again the answer is no. Investigators concluded that the plane had been overloaded and pilot error was to blame for the tragedy. This crash, by the way, was the major news story in Canada during the entire March 1989 solar storm episode and displaced the Quebec blackout to page 3 in the *Toronto Sun*.

During the February 9, 1907, Great Aurora, Atlantic transport liner *Menominae* from Antwerp was struck off Beachy Head on the evening of February 9 by the French steamer *President Leroy Lallier*. The steamer was observed to move erratically in course before collision. During another severe Great Aurora on February 24, 1956, the paper announced that six planes with sixteen missionaries on board were reported missing. The planes had left Cuba en route to the neighboring island of Jamaica at 2:00 P.M. but had never arrived some four hours later. There were no clouds or storms in the area.

These examples show rather dramatically why it is so important to study specific events carefully before jumping to the conclusion that one caused or facilitated the other. It is not sufficient to merely note a coincidence in time and a suspected pathway of impact. Shipwrecks and crashes happen all the time. The odds are very good that they will happen during solar storm disturbances, which are also rather common during any given year. It's just a matter of the odds catching up with you every once in a while.

Magnetic compasses are not the only things that seem to need the magnetic field as a stable reference in time and space. There are many other "systems" on the Earth that need to sense their direction to get to food, shelter, or simply maintain equilibrium in a thousand other ways. The geomagnetic field is so subtle you can't feel its presence outright. But somehow, over millions of years, it seems that organic evolution has managed to detect this force by trial and error and incorporate it into the guidance systems of everything from bacteria to sharks. Even the common monarch butterfly relies on a magnetic sense to orient itself on its annual southward migration to Mexico. Beyond airplanes and ships, there are many other natural systems that could be sensitive to geomagnetic storms, at least in principle. In this particular arena, we have to walk even more carefully among the possible instances of cause and effect.

Back in 1974, Richard Blakemore and Richard Frankel at the University of New Hampshire uncovered a remarkable trick that certain kinds of freshwater bacteria seemed to share. As they grow to maturity, each of them creates within their single-celled bodies nearly two dozen pure cubical crystals of magnetite. Like pearls on a string, the crystals are oriented with the long axis of the bacterium. One can imagine that by some evolutionary process primitive organisms grew a single crystal of magnetite, perhaps as an annoying by-product of eating. As these crystal wastes accumulated, the host became more efficient in finding its way to new locations rather than spinning around and around in the dark. Whatever the process, lowly bacteria during billions of years of evolution managed to beat humans to the discovery of the magnetic compass by, oh, about three billion years!

Using magnetite as a clue, scientists have thrown many different organisms under the microscope, and many have now been found to have at least some kind of magnetite embedded in them, including homing pigeons, tuna, honey bees, dolphins, whales, green turtles, and Elvis Presley. Richard Frankel at the Massachusetts Institute of Technology has gone so far as to herald these discoveries as "the beginning of a new chapter in the story of the interaction of the biosphere with the geomagnetic field." They may have jumped the gun a bit.

The story has become legendary about how homing pigeon rallies are not held during times when geomagnetic conditions are unstable. Since 1980, studies seem to show that pigeons placed blindfolded in a pen and allowed to move tend to move most often in the direction of magnetic north. No single pigeon has been found to do this, but only large numbers of repeated trials turn up this "behavior." Very recently, magnetite has been found in a certain anatomical feature of the heads of pigeons called the ethmoid cavity, which at least looks like a potent argument that they have the hardware needed to fashion a magnetic compass. Although there are investigators excited by this evidence, others are not as convinced that pigeons use the magnetic field. Also in 1998, thousands of pigeons suddenly disappeared during an East Coast race. Some were later recovered in distant farms in Ohio. The unlucky pigeon racers, who lost thousands of dollars in trained birds, blamed geomagnetic storms at first, but there was no evidence that any significant disturbances were going on at the time. Perhaps the pigeons suffered a mild head cold or some other malady that spread rapidly through the flock and disoriented them. We will never know for certain what caused this race to be routed so mysteriously.

Sometimes, however, animal navigation can go awry for reasons having nothing to do with some internal magnetic compass. Dolphins and whales seem to have the required magnetite bodies buried inside their heads, and it is reasonable to wonder whether these mammals navigate shallow coastline waters by taking magnetic bearings. Perhaps the numerous whale and dolphin mass beachings that we hear about may come from geomagnetic storms that disrupt their travels and cause them to head to shore. Although this scenario sounds plausible, it may well be that the tragic beachings we hear about from time to time have an entirely man-made cause. In a recent *Washington Post* article, "Navy Tests Linked to Beaching of Whales," we hear about the plight of seemingly healthy beaked whales in the Bahamas whose mass strandings have now been tentatively traced to U.S. Navy underwater sonar tests and explosions. High-intensity, low-frequency sonar systems emit loud blasts of noise for detecting quiet enemy submarines. These systems also give whales and other marine mammals "terrific headaches" and severe disorientation. A dozen beaked whales also stranded themselves in Greece in 1996 during NATO exercises using similar "active sonar" systems.

But apart from the fact that some animals seem able to do so, exactly how does an organism "sense" which direction magnetite crystals are pointing inside them? How do you know which way a dollar bill is oriented in your pocket? For magnetotaxic bacteria, the magnetic field of the Earth acts on the magnetite crystals to actually turn them into the correct orientation. Bacteria are so light that even dead ones align with magnetic north like the arrow of a compass. Larger organisms, however, are much too big to be physically moved in this way. They need some internal "magnetic sense" that they can recognize, much as we sense our body orientation thanks to the semicircular canals in our middle ear.

The way some organisms might perceive their magnetic surroundings seems to have been discovered since 1978. Microscopic examination of the magnetite crystals detected inside animals as diverse as rodents and humans turn up nervous tissue surrounding these nodules. That many of these organisms are literally "led by the nose" is suggested by the fact that the magnetite concentrations in pigeons, dolphins, whales, tuna, and marlin are found in the ethmoid cavity, located where the bones of the walls and septum of the nasal cavity join. Are humans left out of this exciting new gold rush of evidence for a hidden sixth sense? Apparently not. Since magnetotaxic organisms were discovered, researchers have also found traces of magnetite in human sinuses in much the same anatomical location as for other large animals.

Searching for a magnetite compass among the billions of cells in an organism is far worse than searching for a magnetic needle in an organic haystack. Also, just because you find magnetite (a not especially rare oxide of iron) inside an organism may not make it a workable compass. Nature, it seems, has also found other peculiar uses for magnetite, causing a variety of different cells to stockpile it for other murky purposes than navigation. There could be good biochemical reasons why organisms accumulate magnetite that may have nothing to do with navigation. For instance, clumps of magnetite produce very powerful local magnetic fields that are known to modify chemical reactions. Some researchers suggest that with magnetite nodules inside cells nature has merely discovered another odd way to catalyze biochemical reactions in certain kinds of cells. To be a good compass, magnetite has to be in the shape of a needle or some other elongated structure. A symmetric nodule simply won't do. In some organisms that contain magnetite, however, there is no good evidence that magnetite crystals are aligned in this way. For instance, in one human brain cell out of about fifty thousand, magnetite seems to be clumped, but not into long linear chains as they are in bacteria that use them for guidance. Also these "magnetocytes," as Joseph Kirchvink at the California Institute of Technology calls them, contain magnetite clumps surrounded by "lipid bilayer membranes . . . containing several hundred distinct proteins of unknown function." According to Kirchvink,

> They are definitely not used to detect the geomagnetic field as they do not contain linear chains of crystallographically aligned magnetite crystals as do magnetotactic bacteria, protozoans, migratory fish and birds. At the risk of engaging in speculation, our best guess is that the magnetite crystals are important for biochemistry.

While bacteria seem to use aligned crystals to serve as magnetic compasses, there is growing evidence that other animals may use a more subtle "chemical compass" to do the trick. Magnetite clumps that serve to alter biochemical reactions may cause changes that can be sensed. James Weaver, a biophysicist at the Massachusetts Institute of Technology, has been looking into how such a chemical direction-finding sense might work, at least mathematically. His findings, based on computer modeling, suggests that the subtle chemical changes produced by magnetite-sensitive reactions is enough to cause feeble magnetic signals to be picked out from the normal noisy hubris of a cell's environment. Although no one as yet has confirmed that magnetite can function in this way in more complicated organisms, at least there are some plausible connections between the state of the external field and changes at the cellular level. What remains to be shown is that these biochemical changes actually get promoted into an organism's awareness of orientation. Perhaps, in the future, researchers will inject chemicals that suppress some of these magnetite-catalyzed reactions and test whether the organism "lost their bearings" or not. But perhaps magnetite works not as a transducer of physical orientation but as a modifier of some "psychic compass."

Another intriguing possibility is that, instead of being biochemically important, the magnetite found dispersed in brain tissue may act in some bioelectric fashion. We have all heard that the brain has a complex electromagnetic "hum" with many different cycles going on all at once: alpha waves, beta waves, etc. An entire biofeedback industry has grown up in the last twenty years to help you modify your brainwaves to make you feel better—at least so the claims say. Curiously, many of these cycles are matched in frequency by far more powerful rhythmic changes in the environmental geomagnetic fields. If human brain tissue contains magnetite, but not in a form that can work as a magnetic compass, could it still act in some way to operate as a "psychic compass"? The evidence seems to suggest that these magnetosomes act to catalyse unknown chemical reactions throughout the brain. We also know that imbalances of neurotransmitters throughout the brain have profound impacts on our moods and other mental states. Could the two be related?

We all know that there are "Lies, damn lies, and statistics," but in one curious study statistical evidence seemed to show that the admissions to mental hospitals in New York correlated with what space physicists call the geomagnetic Kp index. It is a measure of how unsettled the geomagnetic field is over the whole planet during a three-hour period. Another study found a similar correlation with the Ap Magnetic Activity index (related to the Kp index) in the psychotic outbursts of patients in a Moscow mental institution. According to Wallace Campbell, of the United States Geological Survey in Denver, Soviet researchers in 1977 reported a correlation of geomagnetic events with the number of heart attacks in Sverdlovsk based on three hundred cases. Even deaths from cardiovascular disease seemed more likely to occur within a day of a geomagnetic storm, as do convulsive seizures and reports of hallucinations. In 1995, Juan Roederer, at the Geophysical Institute of the University of Alaska in Fairbanks, summarized many of these medical studies in an *American Geophysical Union* article, "Are Magnetic Storms Hazardous to Your Health?" Taken together, they did seem to show that something very odd was going on; as he suggests, it would be a good idea to look into the studies more carefully.

Navigation problems, power outages, and communication interference are all symptoms of solar storms changing our environment and causing natural electromagnetic processes to escalate until they become a technological problem. Changing fields and currents find harbor in wires, cause ionospheric changes, and perturb local magnetic fields. Although these impacts seem a bit remote and elusive at times, now that we have entered the Space Age, we have begun to fall victim to far more direct impacts from these same storms.

6 They Call Them "Satellite Anomalies"

> Space weather is working its way into the national consciousness as we see an increasing number of problems with parts of our technological infrastructure such as satellite failures and widespread electrical power brownouts and blackouts.
>
> —National Space Weather Program, "Implementation Plan," 1999

January 20, 1994, was a moderately active day for the Sun. There were no obvious solar flares in progress and there was no evidence for any larger than normal amounts of X rays, but a series of coronal holes had just rotated across the Sun between January 13–19. According to the National Oceanic and Atmospheric Administration's Space Environment Center, the only sign of unrest near the Earth was the high-speed solar wind from these coronal holes, which had produced measurable geomagnetic storm conditions in their wake. NASA's *SAMPEX* satellite was beginning to tell another, more ominous, story. The Sun was quiet, but there were unmistakable signs that energetic electrons were being spawned near geosynchronous orbit, and their concentrations were climbing rapidly. These particles came from the passage of a disturbance from the magnetotail region into the inner magnetic field regions around the Earth. Within minutes, the *GOES-4* and *GOES-5* weather satellites began to detect accumulating electrostatic charges on their outer surfaces. Unlike the discharge you feel after shuffling across a floor, there is no easy and quick way that satellites can unload the excess charges they accumulate, and so they continue to build up until the surfaces reach voltages of hundreds, or even thousands, of volts.

The *Anik E1* and *E2* satellites, owned by Telesat Canada, were a twin pair of GE Astro Space model 5000 satellites, weighing about seven thousand pounds and launched into space in 1991. From their orbital slots on the equator nine hundred miles southwest of Mexico City and fifteen hundred miles apart in space, they soon became the most powerful satellites in commercial use in all of North America. Virtually all of Canada's television broadcast traffic passed through the *E2* transponders at one time or another. The *E2* satellite provided the business community with a variety of voice, data, and image services. Despite some technical difficulties with the deployment of the *Anik E2* antenna, which dogged engineers for several months, the satellites soon became a reliable cornerstone for North American commerce and entertainment.

Canadians eagerly awaited the start of the *Anik-E* satellite service because major cities were few and far between across Canada, a territory bigger than the United States. With hundreds of small towns, and only a few dozen major cities with television stations, the satellites quickly became the information lifeline for many parts of Canada. Twenty-three hundred cable systems throughout Canada and nearly one hundred thousand home satellite dish owners depended on these satellites to receive their programming. Newspapers relied on these satellites to beam their newspapers to distant printing presses to serve far-flung arctic communities. Most people thought the satellites would continue working until at least 2003, but on January 20, 1994, this optimism came to an end.

As the *GOES* satellites began to record electric charges from the influx of energetic particles, the *Intelsat-K* satellite began to wobble on January 20, 1994, and experienced a short outage of service. About two hours later, the *Anik* satellites took their turn in dealing with these changing space conditions and did not do as well. The satellites experienced almost identical failures having to do with their momentum wheel control systems, which help to keep the satellite properly pointed. The first to go was *Anik E1* at 12:40 P.M., when it began to roll end over end uncontrollably. The Canadian press was unable to deliver news to over 100 newspapers and 450 radio stations for the rest of the day but was able to use the Internet as an emergency backup. Telephone users in forty northern Canadian communities were left without service. It took over seven hours for Telesat Canada's engineers to correct *Anik E1*'s pointing problems using a backup momentum wheel system.

About seventy minutes later, at 9:10 P.M., the *Anik E2* satellite's momentum wheel system failed, but its backup system also failed, so the satellite continued to spin slowly, rendering it useless. This time, 3.6 million Canadians were affected as their only source of TV signals went out of service; in an instant, television sets became useless pieces of furniture. Popular programs such as *MuchMusic*, *TSN*, and the *Weather Channel* were knocked off the air for three hours while engineers rerouted the services to *Anik E1*. For many months, Telesat Canada wrestled with the enormous problem of trying to reestablish control of *Anik E2*. They were not about to scrap a $300 million satellite without putting up a fight. After five months of hard work, they were at last able to regain control of *Anik E2* on June 21, 1994. The bad news was that instead of relying on the satellite's disabled pointing system they would send commands up to the satellite to fire its thrusters every minute or so to keep it properly pointed. This ground intervention would have to continue until the satellite ran out of thruster fuel, shortening its lifespan by several years. The good news was that Telesat Canada became the first satellite company to actively stabilize a satellite using "active ground loop control" without using onboard satellite attitude system. In the end, it would turn out to be something of a Pyrrhic victory because on March 26, 1996, at 3:45 P.M., a critical diode on the *Anik E1* solar panel shorted out, causing a permanent loss of half the satellite's power. Investigators later concluded that this, too, was caused by an unlucky solar event.

The connection between the geomagnetic disturbance and the *Anik* satellite outages seemed to be entirely straightforward to the satellite owners at the time, and Telesat Canada publicly acknowledged the cause-and-effect relationship in press releases and news conferences following the outages. They also admitted that the *Anik* space weather disturbance that had ultimately cost their company nearly $5 million to fix was consistent with past spacecraft-affecting events they had noticed and that very similar problems had also bedeviled the *Anik-B* satellite fifteen years earlier. What also made this story interesting is that the *Intelsat-K* and the two *Anik* satellites are of the same satellite design. The key difference, however, is that the Intelsat Corporation specifically modifies its satellites to survive electrostatic disturbances including solar storms and cosmic rays. This allowed the *Intelsat-K* satellite to recover quickly following the storms that disabled the unmodified *Anik* satellites. Clearly, it is possible, and desirable, to "harden" satellite systems so that they are more resistant to solar storm damage. This lesson in spacecraft design is not a new one we have just learned since the *Anik* outage but a very old one that has been applied more or less conscientiously since the dawn of the Space Age itself when these problems were first uncovered.

Although the USSR managed to surprise the United States by orbiting *Sputnik 1*, our entry into the Space Age came in 1958 with the launch of the *Explorer 1* satellite. The main objective of the satellite was simply to staunch the perception that we had fallen behind the USSR in a critical technological area. So the satellite—no bigger than a large beach ball—was put on the engineering fast-track and equipped with a simple experiment devised by James van Allen at the University of Iowa. Even before the first satellite entered the space environment, scientists had long suspected that there would be some interesting things for instruments to measure when they got there, among them, elusive particles called cosmic rays. What they couldn't imagine was that billions of dollars of satellite real estate would eventually fall victim to these same cosmic bullets.

More than ten years earlier, physicists working with photographic films on mountain tops had detected a rainstorm of "cosmic rays" streaming into the atmosphere, but their origins were unknown. Van Allen wanted to measure how intense this rain was before it was muffled by the Earth's blanket of atmosphere, and perhaps even sniff out a clue about where they were coming from in the first place. His experiment was nothing more than a Geiger counter tucked inside the satellite, but no sooner was the satellite in space than the instrument began to register the clicks of incoming energetic particles. Space was indeed "radioactive." Since then, the impact these particles have had on delicate satellite electronics has been well documented by both civilian and military scientists.

Satellites receive their operating power from large-area solar panels that have surfaces covered by solar cells. When the Sun ejects clouds of high-energy protons, these particles can literally scour the surfaces of these solar cells. Direct collisions between the high-speed protons and the atoms of silicon in the cells cause the silicon atoms to violently shift position. These shifting atoms produce crystal defects that increase the resistance of the solar cells to the currents of electricity they are producing. Solar cell efficiency steadily decreases, and so does the power produced by the solar panels. Engineers have learned to compensate for this erosion of power by making solar panels oversized. This lets the satellite start out with extra capacity to cover for this steady degradation of electrical output. But this degradation doesn't happen smoothly over time. Like a sudden summertime hailstorm, the Sun produces unpredictable bursts of particles that do considerable damage in only a few hours.

A series of powerful solar proton events during October 19–26, 1989, for example, caused many satellites to experience severe solar panel degradation in a few days. According to Joe Allen at NOAA, the power output from the solar panels could be carefully followed for GOES satellites, and the October events collectively caused them to lose five years of operating lifetime. This incident also provides an example of how hard it can be to track down accurate information in some space weather impacts. *Aviation Week* and *Space Technology* published an article eight years later in which a report claimed that the GOES-7 satellite itself suffered a five-year, 50 percent mission lifetime loss from this event.

High-energy particles also do considerable internal damage to spacecraft. At the atomic scale, to an incoming proton or electron, the walls of a satellite look more like a porous spaghetti colander than some solid impenetrable wall of matter. High-energy protons can also collide with atoms in the walls of satellite and produce sprays of secondary energetic electrons that penetrate even deeper into the interior of the satellite. Engineers call this "internal dielectric charging." As the batterylike charging of the satellite continues, eventually the electrical properties of some portion of the satellite breaks down and a discharge is produced. In a word, you end up with a miniature lightning bolt that causes a current to flow in some part of an electrical circuit it's not supposed to. As anyone who has inserted new boards into their PC can tell you, just one static discharge can destroy the circuitry on a board. Energetic particles can also deliver their charges directly to the microscopic junctions in electronic circuitry and change information stored in a computer's memory.

Microscopic current flows can flip a computer memory position from "1" to "0" or cause some components, or an entire spacecraft system, to switch on when it shouldn't. When this happens, it is called a "single event upset," or SEU, and like water they come in two flavors: hard and soft. A hard SEU actually does irreparable physical damage to a junction or part of a microcircuit. A soft SEU merely changes a binary value stored in a device's memory, and this can be corrected by simply rebooting the device. Electrostatic discharges can also cause phantom command events. Engineers on the ground cannot watch the circuitry of a satellite as it undergoes an electrostatic discharge or SEU event, but they can monitor the functions of the satellite. When these change suddenly, and without any logical or human cause, they are called "satellite anomalies." They happen a lot more often than you will ever read about in the news media. With hundreds of satellites operating for several decades, over nine thousand anomalies have been recorded by clients of NOAA's National Geophysical Center.

Gordon Wrenn is the retired section leader of the Space and Communications Department of DRA Farnborough in England. Some years ago, he looked into a rash of unexpected changes in an unnamed geosynchronous satellite's pointing direction. The owners of the satellite let him look at their data under the condition that he not divulge its name or who owned it. When the anomalies were compared to the radiation sensor data from the *GOES-7* and *METEOSAT-3* satellites, it was pretty clear that they correlated with increases in the number of energetic electrons detected by *GOES-7*. These insights, however, cannot be uncovered without cooperation from the satellite owners. The specific way that

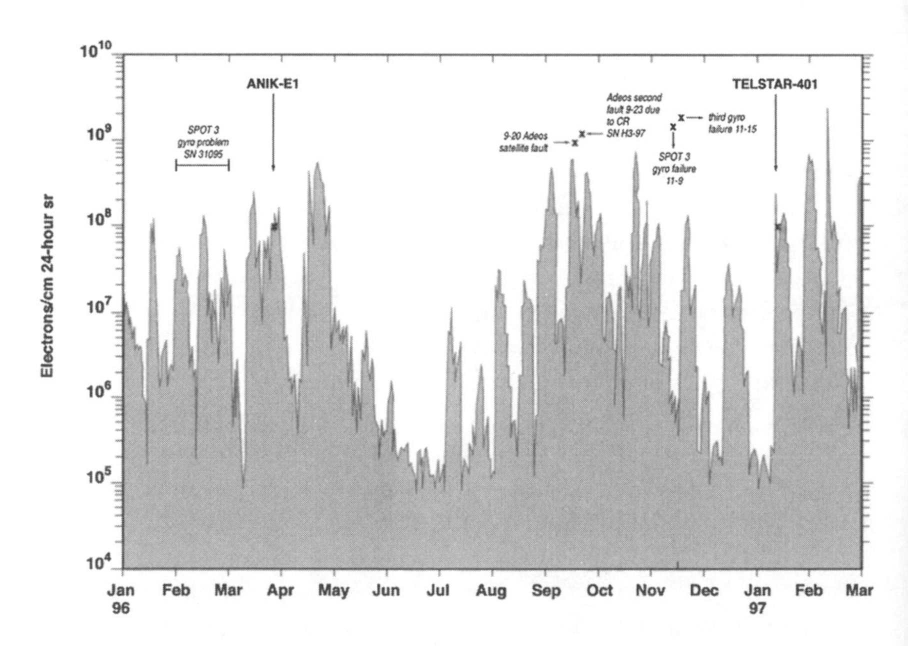

FIGURE 6.1 The rise and fall of energetic particles recorded by the *GOES-9* satellite around the time that *Telstar 401* was disabled in orbit.
source: *Daniel Wilkinson, NOAA/NGDC*

energetic particles cause internal dielectric charging can only be ferreted out when satellite owners provide investigators with satellite data, as Wrenn explains, "Prompt and open reporting offers the opportunity to learn from others' mistakes. Sometimes the lesson can be fairly inexpensive; Telesat Canada were not so fortunate [with the loss of the *Anik* satellites]."

You can get information about satellite anomalies from government research and communication satellites because the information is, at least in principle, open to public scrutiny. The only problem is that you need to know who to talk to, or you have to be willing to comb through hundreds of technical reports, almost none of which are available on the Internet.

The first satellite in the NASA Tracking and Data Relay Satellite System (*TDRSS-1*) was launched in April 1983, and from that time onward the satellite has been continuously affected by soft SEUs. The satellite anomalies affected the spacecraft's attitude control system, and, like mosquitoes on a warm day, they remain a constant problem today. The SEUs have been traced to changes in the computer's RAM, and the most serious of these SEUs were considered mission threatening. If left uncorrected, they could lead to the satellite tumbling out of control. Ground controllers have to constantly keep watch on the satellite's systems to make certain it keeps its antennas pointed in the right direction. This has become such an onerous task that one of the ground controllers, the late Don Vinson, once quipped, "If this [the repeated SEU's] keeps up, TDRS will have to be equipped with a joystick."

The problems with *TDRSS-1* quickly forced NASA to redesign the next satellites in the series, *TDRSS-3* and *4* (*TDRSS-2* was lost in the *Challenger* accident), and the solution was fortunately very simple. In engineering speak, "The Fairchild static, bi-polar 93L422 RAMS were swapped for a radiation-hardened RCA CMM5114 device based on a different semiconductor technology." Radiation hardening is a complex process of redesigning microcircuits so that they are more resistant to the high-energy particles that pass through them. The result is that the two new *TDRSS* satellites have recorded very infrequent SEUs, while, during the same operation period, hundreds still cause *TDRSS-1* to rock and roll, keeping the satellite's human handlers steadily employed for the foreseeable future.

Additional examples of satellites that have suffered from serious damage are harder to find because commercial satellite companies do not want it widely known what the cause of a satellite problem is. The military, on the other hand, also considers this kind of satellite vulnerability information a sensitive issue. Although military satellite impacts are inaccessible, it is possible to seek out, from news reports and a variety of trade journals, many examples of satellite problems caused by, or likely to have been caused by, solar storm events. As with all of the other problems we have seen so far, the biggest lightning rod for these events was the major solar storm during the last solar cycle in March 1989. Over eighty thousand objects are tracked by the powerful radars used by the U.S. Air Force Space Command, but, during this storm, over thirteen hundred of the objects moved from the "identified" to the "unidentified" category as increased atmospheric drag affected their orbits and temporarily converted them into unidentified objects. Later on that same year, another powerful flare between August 15–16 led to a series of geomagnetic events on August 28–29 that caused half the *GEOS-6* telemetry circuits to fail immediately. Meanwhile, back on Earth, the Toronto Stock Exchange closed unexpectedly when all three of their "fault tolerant" disk drives crashed at the same time. This later incident may have been a coincidence—we just don't know for certain.

A particularly common way for satellites to fail is for their attitude control systems to be damaged or compromised in some way. Why this happens has a lot to do with how a satellite recognizes its orientation in space. These systems contain a set of sensors to determine the direction that a satellite is pointing in space, a set of thrusters or gyros to move the satellite in three directions, and a system for "dumping" angular momentum, usually through a mechanical component called a momentum wheel. The basic operating principle for many of these attitude systems is to use some type of sensor or "star tracker" to take frequent images of the sky and compare the locations of the detected stars with an internal catalog. A computer then compares the position differences and causes the satellite to reorient itself to point in the right direction. Energetic particles can impact sensitive electronic camera elements, specifically the so-called CCD chip, and produce false stars. This causes added wear and tear on the entire pointing system as the satellite uses up fuel and the momentum wheel system is needlessly exercised. Even the Hubble Space Telescope, whose mission is actually to observe stars, sees more of these than it is supposed to, because its attitude system and CCD cameras are also under steady attack every day. By the way, it also uses the Fairchild 93LA22 RAM that was employed in TDRSS-1. During September 29, 1989, a strong proton flare caused power panel and star tracker upsets on NASA's *Magellan* spacecraft en route to Venus. The storm was also detected near Earth by the *GEOS-7* satellite. The burst of high-energy protons from the distant sun was the most powerful one recorded since February 1956.

Earlier generations of communications satellites that didn't require star trackers for high-precision pointing used an even simpler position system. Because of the very large transmission beam sizes that cover entire continents, these satellites used magnetometer sensors that detected the local magnetic field of the Earth. Onboard pointing systems compared the detected field orientation against an internal table of what it ought to be if the satellite were pointing correctly. Although using the local magnetic field only gives pointing measurements that are good to a degree or so, this is often good enough for some types of satellites. According to collected reports by Joe Allen, during the March 13–14, 1989, solar storm that triggered the Quebec blackout, the accompanying geomagnetic storm caused many satellite problems. Geosynchronous satellites, which used the Earth's magnetic field to determine their orientation, had to be manually controlled to keep them from trying to flip upside down as the orientation of the magnetic field became disturbed and changed polarity. Records show that some low-altitude, high-inclination, and polar-orbiting satellites experienced uncontrolled tumbling. Even today, the *Iridium* satellite network, for example, also uses magnetometers as a part of their pointing system and so are, at least in principle, potential victims of geomagnetic disturbances.

When a satellite changes its pointing direction, it can either do so by using thrusters or by pushing against an internal mass of some kind. Thrusters are quite messy and only used for gross maneuvers, so satellites use a momentum wheel to provide a countermass to push against as they are turned. A momentum wheel is a symmetric mass of material oriented so that the spin axis is exactly along the major axis of the satellite. Each time the satellite pointing direction is altered slightly, the laws of physics require that each push has to be matched by one in the opposite direction. It is this latter one that causes the momentum wheels to spin up as the satellite pushes in the opposite direction against the momentum wheel to alter its pointing direction. Eventually the rotational energy has to be unloaded or "dumped" so that the momentum wheel system doesn't, literally, fly apart. According to Allen, during October 19–26, 1989, solar storm sequence, an unnamed thirteen-satellite geosynchronous satellite constellation reported 187 "glitches" with its attitude system. Beyond the problem of the attitude control system is the issue of general component vulnerability.

The introduction of off-the-shelf components into the design of satellites has been one of the major revolutions pointed to in recent years by satellite manufacturers, and this keeps space access costs plummeting. It is increasingly being touted as good news for consumers, because the cost-per-satellite becomes very low when items can be mass-produced rather than built one at a time. Based on its experience with the seventy-two-satellite *Iridium* series (now being deorbited), in 2000–2001, Motorola will begin the fourteen-month mass production of the 288 satellites for the Teledesic network in the fastest satellite construction project ever attempted. According to Chris Galvin, CEO for Motorola, their perception is that "satellites are not rocket science so much any more as much as [simply] assembly." This attitude has come to revolutionize the way that satellite manufacturers view their products and estimate the risks of an enterprise.

But there is a downside to this exuberance and economic savings. Most of this revolution in thinking has happened during the mid-1990s while solar activity has been low between the peaks of Solar Cycle 22 and 23. The fact that energetic particles can invade poorly shielded satellites and disrupt sensitive electronics in a variety of ways is not a recently discovered phenomenon that we have to experimentally reconfirm. It has been a fact of life for satellite engineers for over forty years. Data from government research satellites, and weather satellites, convincingly show that the particulate showers from solar wind particles, cosmic rays, solar flares, energetic proton events, and CMEs can all affect spacecraft electronics in a variety of ways. Some of these are inconsequential, others can be fatal. They do not constitute a mystery that we have only encountered by actually placing expensive satellites in harm's way. For this reason, our current situation with respect to solar storms and satellite technology is very different than when previous technologies were developed and deployed for commercial use. It typically took decades for earlier technologies to begin to show signs of sensitivity.

Even more troubling than satellite electronics is that energetic neutrons and other fission fragments produced when solar flare particles strike atoms in the Earth's atmosphere can travel all the way to the ground. There they affect aircraft avionics, causing temporary glitches in both civilian and military aircraft. About one in ten avionics errors are "unconfirmed," which means that no obvious hardware or software problem was ever found to have caused them. In another related incident, an engineer working for American Airlines was curious about a spate of computer glitches that occurred during sales transactions on Trans-Pacific flights. A follow-up investigation by Joe Allen at NOAA confirmed that the glitches matched the record of large magnetic storms or auroral conditions then in progress. This was an exciting result that seemed to confirm that high-flying jet airlines could be directly affected by invisible solar and geomagnetic events. Unfortunately, when the investigation tried to contact the engineer for more data, American Airlines announced that the engineer no longer worked for them. One important source of information on these particles, believe it or not, is cardiac pacemakers. Millions of these are installed in people each year, many of whom take trips on jet planes. They record any irregularities in the rate at which they trigger their pulses, and this information can be examined when their operation logs are return to ground and downloaded by doctors for study. Those glitches recorded among airline staff who wear pacemakers do correlate with solar activity levels. There is also another "down to-earth" problem with these solar storm particles. Whenever computers crash for no apparent reason, some new studies suggest that energetic particles from solar flares may also be to blame. With more components crammed onto smaller chips, the sizes of these components has shrunk to the point where designers are now paying close attention to solar flares. The very popular American Micro Devices K-6 processor, for example, was designed using SEU modeling programs. Because these particles cannot be eliminated by shielding, they may prove the final, ultimate limit to just how small—and how fast—designers can make the next generations of computers.

Even though the tried-and-true approach to reducing radiation effects is to increase the amount of shielding in a satellite, this will not work for all types of radiation encountered in space. For example, the *APEX* satellite investigators concluded, "Conventional shielding is not an effective means to reduce SEUs in space systems that traverse the inner high energy proton belt."

The reason for this is that the particles most effective in producing SEUs are the energetic protons with energies above 40 million volts. When these enter spacecraft shielding, they collide with atoms in the shielding to spawn showers of still more particles. In fact, the thicker the shielding, the more secondary particles are produced to penetrate still deeper into the satellite. Low energy particles, however, can be stopped by nothing more than a quarter-inch of aluminum shielding. For ground-based circuit designers, shielding posseses its own problems because many shielding materials contain naturally occurring radioactive isotopes that produce their own energetic particles. Even the lead in the solder used to make electrical connections poses a severe problem.

For *TDRSS-1*, it was too late to do anything to make the satellite less susceptable to SEUs; however, subsequent satellites in the *TDRSS* series were equipped with radiation-hardened "chips," which virtually eliminated further SEUs in these satellite systems. The pace of developing space-qualified electronics is sluggish at best. Commercial computer systems now operate with 500–1000 megahertz processors and 10 gigabyte memories, but the Space Shuttle was only recently upgraded to an IBM 80386 system. The difference is that the shuttle's "386" can withstand major bursts of radiation and still operate reliably. Intel Corporation and the Department of Defense announced in 1998 that Sandia National Laboratories will receive a license to use the $1 billion Pentium processor design to develop a custom-made radiation-hardened version for U.S. space and defense purposes. The process of developing "rad-hard" versions of current high-performance microchips is complicated because the tricks used to increase chip speed, such as thin wiring and close packing of components, often make the chip vulnerable to ionizing radiation. Larger than commercial etched wiring and thinner than commercial oxide layer deposition are the keys to making chips hardier, it seems. The reason these efforts are expended is pretty simple, though expensive. Peter Winokur, a physicist at Sandia, noted that "when a satellite fails in space, it's hard to send a repair crew to see what broke. You need to put in parts as reliable as possible from the beginning to prevent future problems."

Telegraph, telephone, and radio communications were invented, and brought into commercial use, before it was fully understood that geomagnetic and solar storms could produce disruptions and interference. With satellite technology, we have understood in considerable detail the kind of environment into which we are inserting them so that the resulting radiation effects can be minimized. Their implications for the reliability of satellite services, have been fully anticipated. There are no great mysteries here that beg exploration by using multimillion dollar satellites as high-tech "test particles." Meanwhile, the design of both satellites in space and power systems here on the ground continue to be driven by considerations that have little to do with solar storms and mitigating their impacts.

7 Business as Usual

> I can't help being a little glad that the telegraph companies have had this object lesson. . . . Wireless is affected by certain things which do not hinder the ordinary lines, but in this matter we have the advantage.
>
> —Guglielmo Marconi, 1909

After the *Galaxy IV* satellite ceased operating on May 19, 1997, millions of pager owners woke up to discover a bit later that their high-tech devices had turned into useless pieces of plastic. When they got into their cars and tried to pump gas at the local service station, the pumps rejected their credit cards because they were unable to use the satellite to transmit and receive verification codes. One hundred thousand privately owned satellite dish systems across North America had to be repointed at a cost of one hundred dollars each. In other locales, Yankee ingenuity found a clever work-around to the loss of *Galaxy IV*. The British Broadcasting Company's news program on Houston's KPFT radio station went silent, so the station turned to the Internet to gain access to the program instead. The station didn't want listeners to miss the exciting story about criminals in Bombay who launder their money through the movie industry and were prone to killing directors if the movies bombed at the box office. Meanwhile, Data Transmission Network Corporation lost service to its 160,000 subscribers, costing the company over $6 million. Newspapers and wire services noted that this was the day that the Muzak died, because the *Galaxy IV* also took with it the feed from the Seattle-based music service. Many who previously thought that "elevator music" was annoying realized just how much they actually missed hearing it for the first times in their lives as they rode elevators in silence and shopped in quiet supermarkets. We can mostly survive these kinds of annoyances and even consider them rather amusing in many ways, but the impact of the satellite outage spread into other life-critical corners of our society as well. Hospitals had trouble paging their doctors on Wednesday morning for emergency calls. Potential organ recipients, who had come to rely on this electronic signaling system to alert them to a life-saving operation, did not get paged. In the ensuing weeks, many newspapers including *USA Today* wrote cautionary stories about our overdependence on satellites for critical tasks and services. Even President Clinton ordered a complete evaluation of our vulnerability to high-tech incidents, some of which could be caused by terrorists.

Satellites represent an entirely unique technology that has grown up simultaneously with our understanding of the geospace environment. The first satellites ever launched, such as *Explorer 1*, were specifically designed to *detect* the space environment and measure it. Less than three years later, the first commercial satellite, *Telstar 1*, was pressed into service and began to misbehave as a consequence of space weather effects. There has never existed a time when we did not appreciate how space weather impacts satellite technology.

The technologies of telegraph, telephone, power line systems, and wireless communication went through short learning phases before becoming mature resources that millions of people could count on. Engineers learned how to best deploy them economically while improving their fidelity and coverage. Customers learned what to expect from them in terms of reliability and how to integrate them into their day-to-day lives. History also shows that the pace of this development was slow and methodical. The telephone was invented in 1871, but it took over ninety years before one hundred million people were using them and expecting regular service. The growing radio communication industry had eight million listeners by 1910; one hundred million by 1940. Satellite technology, on the other hand, took less than five years before it impacted one hundred million people between the launch of *Explorer 1* in 1958, and the *Telstar* satellite in 1963. This revolution happened with a single 2-watt receiver on a 170-pound satellite. Even today, wireless cellular telephones have reached over one hundred million consumers in less than five years since their introduction circa 1990. How many times a month do you "swipe" your ATM card at the service station to pump gas? Retail cash verification systems are sweeping the country and all use satellites at some point in their validation process. Have you used a pager or cellular telephone this week? At least some of your connection may be carried by satellite, especially during long-distance calls.

The International Telecommunications Union in Geneva predicted that from 1996 to 2005, the demand for voice and data transmission services would increase from $700 billion to $1.2 trillion. The fraction carried by satellite services will reach a staggering $80 billion. Thomas Watts, vice president of Merrill Lynch's U.S. Fundamental Equity Research conducted a study that predicted $171 billion per year in global revenues by 2007. To meet this demand, many commercial companies are launching aggressive networks of Low Earth Orbit satellites, the new frontier in satellite communications.

In the eyes of the satellite community, we live in a neo-Aristotelian universe. The sublunar realm, as the ancients used to call it, is sectioned into several distinct arenas, each with its own technological challenges and opportunities. Most manned activities involving the Space Shuttle and Space Station take place in orbits from two hundred to five hundred miles. Pound for pound, this is the least expensive environment in which to place a satellite, but it is also useless because with ninety-minute orbital periods, they provide links between points on the globe that last only a few minutes per call.

Then we have low earth orbit (LEO), which spans a zone approximately from four hundred to fifteen hundred miles above the surface. It is the current darling of the satellite industry because from these orbits the round trip time radio signal delays are only about 0.01 seconds compared to the 0.20 seconds delay from geosynchronous orbit. But there are, of course, several liabilities with such low orbits. The biggest of these is the atmosphere of the Earth itself and the way it inflates during solar storm events. This causes high-drag conditions that lower satellite orbits by tens of miles at a time. In addition, the most intense regions of the Van Allen radiation belts reach down to four hundred miles over South America. Satellites in LEO will spend a significant part of their orbital periods flying through these clouds of energetic particles.

Between six thousand and twelve thousand miles, we enter mid earth orbit (MEO) space, which was originally used by the *Telstar 1* satellite but is currently not economically worth the incremental advantage it provides for communication satellites. ICO Global Communications,

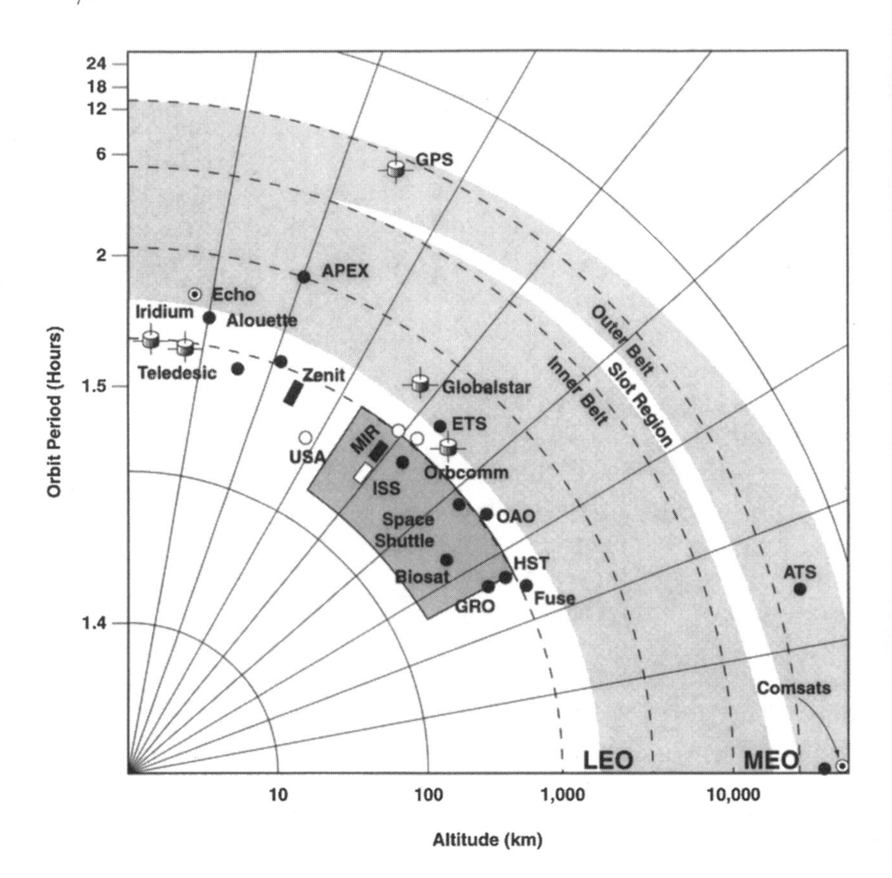

FIGURE 7.1 Location of research and communications satellites compared to the van Allen radiation belts. All geosynchronous satellites are located under the dot on the lower right-hand corner. Shaded area indicates the van Allen belts, including the low-density "slot region" between the inner and outer belts.

renamed "New ICO" following its release from bankruptcy on May 17, 2000, will be setting up a MEO network of twelve satellites costing $4.5 billion at 6,000 miles to provide a global telephone service for an anticipated fourteen million subscribers soon after its completion in 2000. This is also the arena in which the popular Global Positioning System satellites operate.

Finally, at 22,300 miles above the Earth's surface, we enter the so-called Clarke Belt of geostationary orbits (GEO). *Syncom* 3 was the first commercial communications satellite boosted into this orbit, back in 1964, and it was used to televise the Olympic games in Tokyo. Although it is nearly a factor of ten times more expensive per pound to place a satellite at such high altitudes compared to LEO and MEO, there are currently over eight hundred satellites in GEO orbit. About one-fourth are still working and are serving the global needs of commercial and military organizations.

Motorola's $5 billion Iridium network, now consisting of seventy-two satellites that includes four spares, was the first large network to take the LEO stage in 1998, although by 2000 the company went bankrupt. By 1999, Orbital Sciences had also completed a thirty-six-satellite LEO network called Orbcomm. Then came Globalstar Communication's fifty-two-satellite network orbiting at 876 miles, which was completed in December 1999 and built at a cost of $3.5 billion. In the near future, LEO will become ever more crowded with new players offering different services. The Ellipso satellite network will be a twelve-satellite LEO system for Mobile Communications Holdings Inc. to be built for $1.4 billion and placed in service in the year 2000. Alcatel's Skybridge will cost $4.2 billion and consist of sixty-four satellites, with service beginning in 2001–2002. It is expected to have over four hundred million subscribers by 2005. Hughes Electronics will put up SpaceWay for a price of $4.3 billion and be ready for users by 2002. Alcatel plans to launch a network of sixty-three LEO satellites for civil navigation beginning in 2003. Their Global Navigation Satellite System-2 will be operational by 2004 in a joint international partnership between European and non-European partners. The cost is about $636 million and the network will provide 15-foot-position accuracy comparable to the military GPS system, which now provides 20-foot accuracy. Telespazio of Italy's Astrolink system of nine satellites will soon join Loral's Cyberstar network in GEO orbit.

If it gets the financial backing it needs from investors, the most amazing and audacious of these "Big LEO" systems is Microsoft's Teledesic system, which has now partnered with New ICO. For $9 billion, 288 satellites will be placed in LEO and provide 64-megabyte data lines for an Internet in the sky. Motorola will build another spacecraft manufacturing facility to support the production of these satellites, and to capitalize on its experience with the Iridium satellite production process, although they had better outdo Iridium in selling their service. To make Teledesic a reality requires a manufacturing schedule of four satellites per week to meet the network availability goals for 2003. These small but sophisticated 2,000-pound satellites will have electric propulsion, laser communication links, and silicon solar panels.

The total cost of these systems alone represents a hardware investment by the commercial satellite industry of over $35 billion between 1998 and 2004. No one really knows just how vulnerable these complex systems will be to solar storms and flares. Based on past experiences, satellite problems when they do appear will probably vary in a complex way depending on the kind of technology they use and their deployment in space. What is apparent from the public description of these networks is that the satellites being planned share some disturbing characteristics: they are all lightweight, sophisticated, and built at the lowest cost following only a handful of design types replicated dozens and even hundreds of times.

Beyond the commercial pressure to venture into the LEO arena, there is the financial pressure to do so at the lowest possible cost. This has resulted in a new outlook on satellite manufacturing techniques that are quite distinct from the strategies used decades ago. According W. David Thompson, president of Spectrum Astro, "The game has changed from who has 2 million square feet and 500 employees, to who can screw it together fastest."

Satellite systems that once cost $300 million each, now cost $50 million or less. Space hardware, tools, and test equipment are now so plentiful that 80 percent of a satellite can be purchased by just about anyone with enough money, and the rest can be built in-house as needed. The race to meet an ad-hoc schedule set by industry competition to "be there first" has led to some rather amazing tradeoffs that would have shocked older, more seasoned satellite engineers. In the 1960s–1980s, you did all you could to make certain that individual satellites were as robust as possible. Satellites were expensive and one of a kind. Today, a very different mentality has grown up in the industry. For instance, Peter de Selding, a reporter for *Space News*, described how Motorola officials had apparently known since at least 1998 that to keep to their launch schedule they would have to put up with in-orbit failures. Motorola was expecting to loose six Iridium satellites per year even before the first one was launched. At the Fourth International Symposium on Small Satellite Systems on September 16, 1998, Motorola officials said that the company would launch seventy-nine Iridium satellites in just over two years by scrapping launch-site procedures that conventional satellite owners swear by. "Schedule is everything. . . . The product we deliver is not a satellite, it is a constellation," says Suzy Dippe, Motorola's senior launch operations manager. A 10 percent satellite failure rate would be tolerated if it meant keeping to a launch schedule. Of course, it is the satellite insurance industry that takes the biggest loss in this failure, at least over the short term.

Satellite insurers are now getting worried that satellite manufacturers may be cutting too many corners in satellite design. The sparring between insurers and manufacturers has become increasingly vocal since 1997. The nearly $600 million in in-orbit satellite failures that insurance companies have had to pay on in 1998 alone has prompted questions whether spacecraft builders are cutting costs in some important way to increase profit margins, especially with the number of satellite anomalies continuing to rise. Between 1995 and 1997, insurance companies paid out 38 percent of the $900 million in claims just for on-orbit satellite difficulties. Since the early 1980s, satellite failure claims have doubled in number, from $200 to $400 million annually. This trend has prompted Benito Pugnanel, deputy general manager at Assicurazioni Generali S.p.A, to lament that "the number of anomalies in satellites appear to be constantly on the rise . . . because prelaunch tests and in-flight qualification of new satellite components have been cut down to reduce costs and sharpen a competitive edge."

But Jack Juraco, senior vice president of commercial programs at Hughes Space and Communications, disagreed rather strenuously with the idea that pre-launch quality was being short-changed. "The number of failures has not gone up as a percent of the total. . . . What has happened is that we have more satellites being launched than previously. The total number of anomalies will go up even if the rate of failure is not increasing."

Hughes Space and Communications, for example, has 67 satellites, and there has been no percentage change in the failure rate. They use this to support the idea that the problems with satellite failures are inherent to the technology, not the satellite environment that changes with the solar cycle. According to Michael Houterman, president of Hughes Space and Communications International, Inc., of Los Angeles, the spate of failures in the *HS-601* satellites is a result of "design defects," not of production schedule pressure or poor workmanship: "Most of our quality problems can be traced back to component design defects. We need, and are working toward, more discipline in our design process so that we can ensure higher rates [of reliability]."

Satellite analyst Timothy Logue at the Washington law firm of Coudert Brothers begs to differ: "The commercial satellite manufacturing industry went to a better, faster, cheaper approach, and it looks like reliability has suffered a bit, at least in the short term."

Per Englesson, deputy manager of Skandia International Stockholm, a satellite insurance company, was rather less impressed by industry's claims that quality was not a part of the problem bedeviling the satellite industry: "Anomalies aboard orbiting satellites have reached unprecedented proportions." Could it be that one reason for the on-board failures is the fact that relatively small organizations are buying satellites but not hiring the technical expertise needed to oversee their construction? He criticized satellite owners who refuse to get involved in technical evaluations of hardware reliability, instead leaving all such issues for insurers to figure out.

It used to be that satellite components, like grapes for a wine, were hand-selected from only the finest and best parts. The term *mil-spec* (military specifications) represented components designed of the highest quality and, in most cases, considerable radiation tolerance. Not anymore. One of the most serious problems that seems to come up again and again is the issue of off-the-shelf electronics. They are readily available, cheap, and are an irresistible lure for satellite manufacturers working within fixed or diminishing budgets. This is frequently touted as good news for consumers because the cost per satellite becomes very low when commercially available analogues to expensive mil-spec components can be used instead. It is, however, not entirely the fault of the satellite designers that nonmil-spec parts are not used. Very often, no mil-spec equivalents are available. Over the last few decades, research and development of mil-spec analogues by commercial manufacturers has been quietly phased out as military needs have changed.

One satellite network that expects to keep costs down by using off-the-shelf electronics is the Teledesic system. But they are already off to what appears to be a rocky start. Reporter Keith Stein for the journal *International Space Industry Report* of May 7, 1998, describes how the *Teledesic 1 (T1)* experimental satellite was no longer operating as planned. Its purpose was to conduct a series of communications tests on three channels between 18–450 MHz to demonstrate the high data rate capabilities of the telemetry that will be used with the full network of 288 satellites. No details were given, either to the cause of the malfunction, the systems involved, or the time when the satellite failed. Meanwhile, Motorola still expects to complete the manufacture of these satellites in a whirlwind fourteen months, just as soon as they get the green light to start production. Celso Azevedo, president and CEO of the Lockheed-Martin-supported Astrolink, has strong opinions about the rush into LEO orbit and the quality of commercial satellite technology. In an interview with *Satellite Communications* magazine he spells out what Teledesic ought to be considering in the balance: "You have to minimize your technological risks. GEO architecture is proven. When utilizing new technology, developers have a tendency to go too far and stretch the envelope, which is what Teledesic is doing. The project is unlaunchable, unfinanceable, and unbuildable."

The number of basic satellite designs also continues to fall as mass production floods geospace with a numerically smaller diversity of satellites based on similar designs and assumptions about space weather hazards by Lockheed-Martin, Hughes, and Motorola. For example, there are currently forty *HS-601* satellites of the same model as *Galaxy IV* in operation, and these include PanAmSat's *Galaxy-7* and DirecTV's *DBS-1* satellites, which also experienced primary control processor failures as did *Galaxy IV*. Motorola's Iridium satellite network lost seven of its identical satellites by August 1998. According to Alden Richards, CEO of a Greenwich, Connecticut risk management firm, "These problems are not insignificant. Insurers are clearly concerned that there have been these anomalies."

Despite a miserable 1998, which cost them over $600 million in in-orbit satellite payouts, insurance companies still regard the risks of in-orbit satellite failures as a manageable problem. Launch services and space insurance markets generated $8 billion in 1997 and $10 billion in 1998. Since private insurers entered the space business in the 1960s, they have collected $4.2 billion in premiums and paid out $3.4 billion in claims. This is a very slim $800 million profit, but at 19 percent it is still considered a good long-term return on investment. Insurers consider today's conditions a buyers' market with $1.2 billion capacity for each $200 million satellite. There is a lot of capacity available to cover risk needs.

Like any insurance policy the average home owner tries to get, it's necessary to deal with a broker and negotiate a package of coverages. In low-risk areas, you pay a low annual premium, but you can pay higher premiums if you are a poor driver, live on an earthquake fault, or own beach property subject to hurricane flooding. In the satellite business, just about every aspect of manufacturing, launching, and operating a satellite can be insured, at rates that depend on the level of riskiness. Typically, for a given satellite, ten to fifteen large insurers (called underwriters) and twenty to thirty smaller ones may participate. There are about thirteen international insurance underwriters that provide about 75 percent or so of the total annual capacity. Usually the satellite insurance premiums are from 8–15 percent for risks associated with the launch itself. In-orbit policies tend to be about 1.2 to 1.5 percent per year for a planned ten to fifteen-year life span once a satellite survives its shakeout period. If a satellite experiences environmental or technological problems in orbit during the initial shakeout period, the insurance premium paid by the satellite owner can jump from 3.5 to 3.7 percent for the duration of the satellite's lifetime. This is the only avenue that insurers have currently agreed upon to protect themselves against the possibility of a complete satellite failure. Once an insurance policy is negotiated, the only way that an insurer can avoid paying out on the full cost of the satellite is in the event of war, a nuclear detonation, confiscation, electromagnetic interference, or willful acts by the satellite owner that jeopardize the satellite. There is no provision for acts of God such as solar storms or other environmental problems. Insurers assume that if a satellite is sensitive to space weather effects, this will show up in the reliability of past models of the satellite (*HS-601*, etc.), which would then cause the insurer to invoke the higher premium rates during the remaining life of the satellite. Insurers, currently, do not pay any attention to the solar cycle but only assess risk based on the past history of the satellite's technology. When that past history extends barely one solar activity cycle, it is easy to become misled about reliability.

As you can well imagine, the relationship between underwriters and the satellite industry is both complicated and at times volatile. Most of the time it can be characterized as cooperative because of the mutual interdependencies between underwriters and satellite owners. During bad years, like 1998, underwriters can lose their hats and make hardly any profit from this calculated risk taking. Over the long term, however, satellite insurance can be a stable source of revenue and profit, especially when the portion of their risk due to launch mishaps is factored out of the equation. As the Cox Report notes about all of this,

> The satellite owner has every incentive to place the satellite in orbit and make it operational because obtaining an insurance settlement in the event of a loss does not help the owner continue to operate

its telecommunications business in the future. To increase the client's motivation to complete the project successfully, underwriters will also ask the client to retail a percentage [typically 20 percent] of the risk. (Cox Report, 1999)

According to Philippe-Alain Duflot, director of the Commercial Division of AGF France,

The main space insurance players have built up long-term relations of trust with the main space industry players, which is to say the launch service providers, satellite manufacturers and operators. And these sustained relations are not going to be called into question on the account of a accident or series of unfortunate incidents.

Still, there are heated disputes that emerge between underwriters and owners, which are now leading to significant changes in this relationship. Satellite owners, for instance, sometimes claim a complete loss on a satellite after it reaches orbit, even if a sizable fraction of its operating capacity remains intact after a "glitch." According to Peter D. Nesgos, of the New York law firm Winthrop, Stimson, Putnam, and Roberts, as quoted by *Space News*, "In more than a dozen recent cases, anomalies have occurred on satellites whose operators say they can no longer fulfill their business plans, even though part of the satellite's capacity can still be used."

This has caused insurance brokers to rethink how they write their policies and insurance underwriters to insist on provisions for partial salvage of the satellite. In 1995, the *Koreasat-1* telecommunications satellite owned by Korea Telecom of South Korea triggered just such a dispute. In a more recent case, underwriters actually sued a satellite manufacturer, Spar Aerospace of Mississauga, Canada, over the *AMSC-1* satellite, demanding a full reimbursement of $135 million. They allege that the manufacturer "covered up test data that showed a Spar-built component was defective." Some insurers are beginning to balk at vague language that seemingly gives satellite owners a blank check to force underwriters to insure just about anything the owners wish to insist on.

One obvious reason satellite owners are openly averse to admitting that space weather is a factor is that it can jeopardize reliability estimates for their technology and thus impact the negotiation between owner and underwriter. If the underwriter deems a satellite poorly designed to mitigate against radiation damage or other impulsive space weather events, they may elect to levy a higher premium rate during the in-orbit phase of the policy. This can cost millions of dollars out of the profit margin. They might also offer a "launch plus five year" rather than a "launch plus one year" shakeout period. This issue is becoming a volatile one. A growing number of stories in the trade journals since 1997 report that insurance companies are increasingly vexed by what they see as a decline in manufacturing techniques and quality control. In a rush to make satellites lighter and more sophisticated, owners such as Iridium LLC were willing to lose six satellites per year. What usually isn't mentioned is that they also request payment from their satellite insurance policy on these losses, and the underwriters then have to pay out tens of millions of dollars per satellite. In essence, the underwriter is forced to pay the owner for using risky satellite designs, even though this works against the whole idea of an underwriter charging higher rates for known risk factors. Of course, when the terms of the policy are negotiated, underwriters are fully aware of this planned risk and failure rate but willing to accept it in order to profit from the other less risky elements of the agreement. It is hard to turn down a five-year policy on a $4 billion network that will only cost them a few hundred million in eventual payouts. The fact is that insurers will insure just about anything that commercial satellite owners can put in orbit, so long as the owners are willing to pay the higher premiums. Space weather enters the equation because, at least publicly, it is a wild card that underwriters have not fully taken into consideration. They seemingly charge the same in-orbit rates (1.2 to 3.7 percent) regardless of which portion of the solar cycle we are in.

More and more often, satellite insurance companies are finding themselves in the position of paying out claims, but not for the very familiar risk of launching the satellite with a particular rocket. In the past, the biggest liability was in launch vehicle failures, not in satellite technology. As more satellites have been placed in orbit successfully, a new body of insurance claims has also grown at an unexpected rate. Jeffrey Cassidy, senior vice president of the aerospace division of A.C.E. Insurance Company Ltd., notes that as many as eleven satellites during 1996 have had insured losses during their first year of operation. The identities of these satellites, or the names of their owners, were not divulged.

According to *Space News*, satellite insurance companies are reeling from the huge payouts of insurance claims that totaled $750 million during the first half of 1998 alone. Most of these claims were for rocket explosions such as the launch of the *Galaxy 10* satellite, which cost $250 million. The remaining claims, however, included in-orbit failures of the *Galaxy IV* ($200 million) and twelve of the Iridium satellites ($254 million). By 1999 a new trend in insurance payouts had begun to emerge. "Satellite Failures Put Big Squeeze on Underwriters," read an article by Peter de Selding in *Space News*: "[1998] will go down in space-insurance logbooks as the most costly in history. . . . One notable trend in 1998 was the fact that failures of satellites already in orbit accounted for more losses than those stemming from rocket failures."

Despite the rough times that both manufacturers and insurers seem to be having, they are both grimly determined to continue their investments. Assicurazioni Generali, S.p.A of Trieste, the world's biggest satellite insurance underwriter, has no plans to reduce its participation in space coverage, but at the same time thinks very poorly of the satellite manufacturing process itself. Giovanni Cobbo, Assicurazioni's space department manager, is quoted as saying, "I would not buy a household appliance that had as many reliability problems as today's satellites." Despite all the dramatic failures, the satellite insurance underwriters have actually lowered their insurance rates for launches from 15–16 percent in 1996 to 12–13 percent in 1997. Meanwhile, in-orbit insurance rates after the shakeout phase have remained at 1–2 percent per year of the total replacement cost. Industry insiders do not expect this pricing to remain so inexpensive. With more satellite failures expected in the next few years, some space industry analysts are convinced that these rates may increase dramatically.

When satellites fail, more often than not another turn of affairs also rings true. The lessons learned from satellite malfunctions beginning with *Telstar 1* seem now to have been publicly lost in the discussions of cause and effect. We have entered a new age when "mysterious" satellite anomalies have suddenly bloomed as if out of nowhere. The default explanation for satellite problems has moved away from public discussions of sensitive technology in a hostile environment to guarded postmortems that point the finger to sometimes obscure technological causes.

Curiously absent from virtually every communications satellite report of a problem is the simple acknowledgment that space is not a benign environment for satellites. Satellite manufacturers often look for technological problems to explain why satellites fail, while scientists look at the spacecraft's environment in space to find triggering events. What seems to be frustrating to the satellite manufacturing industry is that, when in-orbit malfunctions occur, each one seems to be unique. The manufacturers can find no obvious pattern to them. Like a tornado entering a trailer park, when space weather effects present themselves in complex ways across a trillion cubic miles of space some satellites can be affected while others remain intact.

These complex systems seem to be remarkably robust, although for many of them, in the wrong place at the wrong time, their failure in orbit can be tied to solar storm events. The data, however, are sparse and circumstantial because we can never retrieve the satellites to determine what actually affected them. The exception is the NASA *LDEF Satellite*, which was retrieved by the shuttle. Currently, we can only speculate that "storm A killed satellite B" or that "a bad switch design was at fault." Since there is no free flow of information between industry, the military, and scientists, and the satellites can't be recovered, the search for a "true cause" remains a maddeningly elusive goal. But the playing field is not exactly level when it comes to scientists and industries searching for answers. This usually works to the direct benefit of the satellite owners.

The scientific position is that we really, truly, don't know for certain why specific satellites fail, no matter how much circumstantial evidence we accumulate. This is especially true when commercial and military satellite owners refuse to tell you the details of how their satellites were affected. The Federal Aviation Authority, without the proverbial black box, would have an awful time recreating the details of plane crashes if they were as data-starved as your typical space scientist or engineer. Another reason for this uncertainty is that things that happen at the same time aren't always related to each other in terms of cause and effect. Sometimes, complex technology does, simply, stop working on its own.

You might recall that at one point it looked very convincing that the *Exxon Valdez* may have had unseen navigation problems caused by a solar storm then in progress. The only problem is that the navigation aids in use on the *Exxon Valdez* are rather immune from magnetic storms, so the plausible story that a magnetic storm caused the Exxon Valdez to be on the wrong course is, itself, quite wrong. This is why it seems to be very hard to tie specific satellite failures to solar storm events, even though from the available circumstantial evidence it looks like a sure bet. You can't recover the satellite to autopsy it and confirm what really happened.

If there are thousands of working satellites in space, why is it that a specific storm seems to affect only a few of them? If solar storms are so potent, why don't they take out many satellites at a time? Solar storms are at least as complex as tornadoes. We know that tornadoes can destroy one house and leave its neighbors unscathed, but this doesn't force us to believe that tornadoes are not involved in the specific damage we see. The problem with solar storms is that they are nearly invisible, and research satellites cannot give us a complete picture of what the storms are doing. We hardly see them coming, and the data to determine specific cause-and-effect relationships is usually incomplete, classified, scattered among hundreds of different institutions, or anecdotal. For this reason, any scientist attempting to correlate a satellite outage or "anomaly" with the outcome of a particular solar storm fights something of an uphill battle. There is usually only circumstantial evidence available, and the details of the satellite design and functions up to the moment of blackout are shrouded in industrial or military secrecy.

Commercial satellite companies, meanwhile, would prefer that this subject not be brought into the light for fear of compromising their fragile competitive edges in a highly competitive market. It isn't because they are afraid of acts of God denying them an insurance payout, because insurance underwriters would cover even solar storm damage. In a volatile industry driven by stock values and quarterly profits, no company wants to tell about their anomalies or make their data public for scientists to study. For example, Iridium's stock took a major tumble during the summer of 1998 in the aftermath of seven satellite outages, as investors got cold feet over the technology. The company eventually went bankrupt in 1999. Also, if a specific satellite design is shown to be vulnerable to space weather, then the next time a similar satellite is flown the insurance underwriters could insist on a higher in-orbit rate (e.g., 3.8 percent), which would cost the owner millions of dollars in annual premiums.

Long before the advent of satellite insurance, the first satellite to fall victim to space weather effects was, in fact, one of the first commercial satellites ever launched into orbit in July 1962: *Telstar 1*. In November of that year, it suddenly ceased to operate. From the data returned by the satellite, Bell Telephone Laboratory engineers on the ground tested a working twin to *Telstar* by subjecting it to artificial radiation sources and were able to get it to fail in the same way. The problem was traced to a single transistor in the satellite's command decoder. Excess charge had accumulated at one of the gates of the transistor, and the remedy was to simply turn off the satellite for a few seconds so the charge could dissipate. This, in fact, did work, and the satellite was brought back into operation in January 1963. The source of this information was not some obscure technical report or an anecdote casually dropped in a conversation. This example of energetic particles in space causing a commercial satellite outage was so uncontroversial at the time that it appeared under the heading "Telstar" in the 1963 edition of the *World Book Encyclopedia*'s *1963 Yearbook*. The satellite was not insured and there was nothing to be gained or lost in concealing the cause of the problem.

Recast in today's polarized atmosphere, the outcome would have been very different. The satellite owner would have declared the failure a problem with a known component of the satellite and promised to fix this design problem in all future satellites. The scientists, meanwhile, would have suspected that it was, instead, a space weather event that had charged the satellite. These findings would be published in obscure journals and trade magazines or appeared as viewgraphs used in technical or scientific presentations. An artificial public "mystery" about why the component failed at that particular moment would have been created, adding to a growing fog of false confusion about why satellites fail in orbit. Despite our increased understanding of space weather effects, more satellites seem to mysteriously succumb to outages while in service. It's as if we are having to learn, all over again, that space is fundamentally a hostile environment, even when it looks benign on the basis of sparse scientific data and the absence of satellite anomalies publicly admitted.

Of course, satellite owners, and, for that matter, electrical utility managers, are unwilling or in some cases unable to itemize every system anomaly, just as major car manufacturers are not about to publicly list all the known defects in their products. On the other hand, industry has quickly accepted the fact that it is cheaper to admit to the rare but significant life-threatening problems and voluntarily recall a product than to wait for a crushing class-action law suit. Rarely do commercial satellite owners give specific dates and times for their outages, and, in the case of Iridium, even the specific satellite designations are suppressed, as is any public discussion about the causes of the outages themselves. If this is to be the wave of the future in commercial satellite reportage, especially from the Big LEO networks, we are in for a protracted period of confusion about causes and effects. Anecdotal information provided by confidential sources will be our only, albeit imperfect, portal into what is going on in the commercial satellite industry. Without specific dates and reasons for failure, scientists cannot then work through the research data and identify plausible space weather effects, or at least demonstrate that they were irrelevant. This also means that the open investigation into why satellites fail, which could lead to improvements in satellite design and improved consumer satisfaction with satellite services, is all but ended. As Robert Sheldon, a space scientist at Boston University's Center for Space Studies, notes,

> The official AT&T failure report [about the *Telstar 401*] as presented by Dr. Lou Lanzerotti at the Spring AGU Meeting denied all space weather influence and instead listed three possible [technological] mechanisms. . . . This denial of space weather influence was met with a murmuring wave of disbelief from the audience who no doubt had vested interests in space weather."

For years, Joe Allen and Daniel Wilkinson at NOAA's Space Environment Center kept a master file of reported satellite anomalies from commercial and military sources. Their collection included well over nine thousand incidents reported up until the 1990s. This voluntary flow of information dried up rather suddenly in 1998 as one satellite owner after another stopped providing these reports or as friendly personal contacts retired. From then on, access to information about satellite problems during Cycle 23 would be nearly impossible to obtain for scientific research. More than ever, examples of satellite problems would have to come from the occasional reports in the open trade literature, and these would only cover the most severe, and infrequent, full outages. There would be no easy record of the far more numerous daily and weekly mishaps, which had been the pattern implied by the frequency of these anomalies in the past.

Meanwhile, the satellite industry seems emboldened by what appears on the surface to be a good record in surviving most solar storm events during the last decade. With billions of dollars of potential revenue to be harvested in the next five to ten years, we will not see an end to the present faceoffs between owners, insurers, and scientists. For the consumer and user of the new satellite-based products: Caveat emptor. The next outage may, however, come as suddenly as a power blackout and find you as ill-prepared to weather its consequences. In the end, solar storms may seek you out in unexpected places and occasions and touch you electronically through your pagers, cellular phones, and Internet connections. All this from across ninety-three million miles of space.

8 Human Factors

> By the time Claggett and Linley reached their [lunar] rover and turned it around, they no longer bothered with their dosimeters, because once the reading passed the 1,000-Rem mark, any further data were irrelevant. They were in trouble and they knew it.
>
> —James Michener, *Space*, 1982

On June 4, 1989, a powerful gas line explosion demolished a section of the 1,153-mile Trans-Siberian Railroad, engulfing two passenger trains in flames. Rescue workers worked frantically to aid the passengers, but only 723 could be saved. The rest perished. Many of the 500 victims were children bound for holiday camps by the Black Sea. "My sister and my aunt are somewhere here in these ashes," said Natalya Khovanska as she stumbled between the remains of the trains, which were still smoldering. The explosion was estimated to have been equal to ten thousand tons of TNT, and it felled all the trees within 3 miles of the blast. By some accounts, a wall of flame nearly 2 miles wide engulfed the valley, hurling twenty-eight railway cars off the tracks. The explosion instantly cut the Soviet Union's gas supply by 20 percent. A commission was quickly set up to investigate the blast, but several days later they had still not determined why it happened, except that pipeline engineers had increased the pressure in the line rather than investigate the sudden pressure drop caused by the leak. The gas from the leak settled into a valley near the towns of Ufa and Asha and the passing trains detonated the lethal mixture. Mikhail Gorbachev denounced the accident as an example of "irresponsibility, incompetence and mismanagement" in an address to the Congress. He even suspected sabotage. The cause of the explosion was later identified as the profound disrepair of the pipeline, which had become badly corroded over time and never properly maintained. In the Urals, the weakened walls had finally given way to the pressure of the gas and begun to breach. Corrosion is a process that geomagnetic storms can have a hand in producing, given the right conditions and a lack of maintenance.

Just as geomagnetic storms can cause currents to flow in telegraph lines and trans-Atlantic cables, under certain circumstances, they can also flow in natural gas pipelines. The Ural pipeline disaster was, by all accounts, an extreme event. Because the pipeline is not oriented in a favorable direction to easily pick up GICs, and because it is, in fact, very far away from the latitudes where GICs are most intense, it is unlikely that geomagnetic activity acting over time had much to do with this disaster. The Alaskan pipeline, on the other hand, extends over eight hundred miles in a north-south direction, and its central third runs along the latitude of the auroral electrojet current. It was built during the 1970s and specifically designed to minimize these currents. Modern pipelines are protected from long-term current flows by a weak countercurrent of a few amperes, which is applied so that the pipeline has a net negative potential relative to ground. The problem is that auroral currents change polarity in minutes, rendering this "cathodic protection" useless. During geomagnetic storms, when the electrojet current flows erratically, currents as high as 1,000 amperes have been detected. The lifetime of the Alaskan pipeline is now estimated to be many years shorter than originally planned. At that time, perhaps a decade from now, we will undoubtedly hear more about aggressive last-ditch countermeasures being employed to plug leaks or replace whole sections of the pipeline. Some of these problems may arrive sooner than later. In 1990, there were well-publicized plans to increase the pressure in the Alaskan pipeline. These plans had to await the results from a detailed government investigation of the pipeline's corrosion. Although investigators turned up evidence of gross negligence on the part of the pipeline inspectors, they gave the project a clean bill of health and allowed the higher pressures to be used. Meanwhile, pipeline engineers in Finland have been monitoring GIC currents in their lines for over a decade and are far more concerned about what the future may bring. According to a report on space weather impacts by the French national space agency CNES, the long-term impacts of these currents can be substantial. Pipelines designed to last fifty years can suffer wall erosion of 10 percent in only fifteen years unless the pipeline is regularly monitored and upgraded. No one seriously expects another devastating explosion such as the one in the Urals from any currently active pipeline. At worst, GICs will enhance the rate of corrosion in certain pipelines in high-latitude countries that will require careful inspection. But there are other situations where human health can be more directly impacted by solar storms.

At 1:20 A.M. EDT on August 4, 1972, the Sun let loose one of the most powerful blasts of radiation ever recorded during the Space Age. Streams of X rays and high-energy protons flowed past the Earth within minutes, but not before triggering a major geomagnetic disturbance that disrupted telephone service and destroyed a power transformer at the British Columbia Hydro and Power Authority. Although ground-based observers were kept on their toes by the unexpected power and communication outages, the event would have had a much more deadly outcome had it arrived four months later, between December 7–19, while Apollo 17 astronauts were outside their spacecraft. Within a few hours, some estimates suggest that Harrison Schmidt and Eugene Cernan would have been hit by an incredible blast of radiation well over 1,000 rem.

The astronauts would have suffered acute radiation sickness by the time they reached their lunar module and probably even died some time later back on Earth. This is why James Michener, in his book *Space*, dwells on a similar, hypothetical, event in his story of the fictional Apollo 18 mission. Some experts downplay what the "Apollo 17" flare might actually have done. Gordon Woodcock, for example, writes in his book, *Space Stations and Platforms*, that

> had an Apollo crew been on the lunar surface during the 1972 flare, they would very likely have received enough radiation to become ill. Radiation sickness effects at an exposure level of a few hundreds of rem take hours or days to become debilitating. James Michener's description in Space was not accurate.

Others beg to differ. According to Alan Tribble's *The Space Environment: Implications for Spacecraft Design*, "During August 1972 and again in October 1989, there were two extremely large solar proton [releases]. If an astronaut had been on the Moon, shielded by just a space suit, the radiation dose would probably have been lethal."

The orbiting command module would not have altered the outcome significantly, according to shielding calculations by physicist Lawrence Townsend and his collaborators at NASA's Langley Research Center. Their "worst-case" analysis shows how the August 1972 solar proton releases would have punched through bulkheads similar to those in the Apollo mission and given the astronauts dosages as high as 250 rems. "Such an acute exposure would likely have incapacitated the crew because of radiation sickness and could possibly be lethal."

Even this dosage is nothing to be sanguine about. Most radiation dosage tables say that 20 percent of the people exposed to even this level are sure to die within a month or two.

Radiation: most of us have an instinctive fear of it. Even the word itself is cloaked in mystery and a sense of foreboding. In reality, we are all more familiar with radiation than we suspect. No matter where you live, you receive fifteen to twenty chest X rays each year of environmental radiation—and there is almost nothing you can do about it. Even at the Earth's surface, under a thick blanket of atmosphere, solar storms add their share to this cargo of potential damage. To see just how this happens, we are going to have to look a bit more quantitatively at what radiation is all about.

As you sit reading this book, you are being pummeled by various forms of electromagnetic energy, from visible light to radio waves. You are also being struck by the sons and daughters of particles that have streamed, literally, from the far corners of our universe. In casual conversation, these kinds of energy are simply called "radiation," even though physicists have known for over a century that their various forms are quite different. Electromagnetic radiation includes the familiar rainbow of the visible spectrum, crammed between a vast range of other waves traveling at the speed of light. Some of these can be stopped by a sheet of ordinary paper. Other more energetic forms of light, like X rays and gamma rays, require ever increasing thicknesses of matter to abate them.

In a separate category of radiation we have fast-moving particles that also come in several basic types such as electrons, protons, and the nuclei of the elements heavier than hydrogen. The amount of damage that these material forms of radiation can inflict depends on how much energy each particle is carrying. The bigger the energy, the more punch they can deliver and the more collateral damage they produce as they penetrate the skin of a spacecraft or the tissues of an organism. Electromagnetic radiation in the ultraviolet can give you a sunburn, but energetic particles can bore their way into your cells and explode like a small bomb, "nuking" a gene.

Just as we can measure temperature in terms of "degrees," it shouldn't surprise you that we can also measure the impact that radiation makes: scientists call it a "rad." When a specific form of radiation delivers one watt of energy into 100 kilograms of tissue, this is one rad. Not all radiation affects tissue equally, so health physicists prefer to use another unit, the rem, to give an actual dosage equivalent for the different types of radiation as they damage biological tissue. For example, in one second, one watt of alpha particles (stripped helium atoms ejected by the decay of heavier radioactive atoms) causes twenty times more damage than absorbing the one watt of X rays or gamma rays. So, for one rad of absorbed dose, you get exactly one rem of equivalent dose, if you are talking about X rays, and 20 rems if you are talking about the more destructive alpha particles. You definitely want to stay away from alpha particles!

Now, how much radiation is too much? Unlike vitamins and money, more radiation is probably not better. Since the start of the cold war and the first nuclear bomb tests, the general public has heard a lot about radiation effects; Hiroshima victims with their skin melting from their bones, genetic mutations, cancer. It is all ghastly stuff, and it is not hard to excuse the image most people have of radiation: that it is always a bad thing. Like many things in nature, radiation is bad in degrees. But unlike the rather obvious summer monsoons that can kill thousands of people at a time, radiation is a stealthy phenomenon that we have learned about only in the last one hundred years of human history. Curiously, for the last few *billion* years, it is a phenomenon that is well known to evolution on this planet. Biologically, even at the cellular level, there are powerful mechanisms at work that can repair most radiation damage to an organism. Man-made forms of radiation, however, tend to be more powerful and concentrated than anything evolution has ever prepared us to deal with. Let's have a look at table 8.1. When you review these numbers, you might want to consider that a typical chest X ray is worth a trifling 0.020 rads (for X rays, remember that one rad is the same as one rem) on the same scale.

The table is appropriate for what will happen during an acute, short-term (minutes to hours) radiation exposure. But, amazingly, if you took

TABLE 8.1 Radiation Dosages in Rads and Their Health Effects

0–50	No obvious effects
80–120	10% chance of vomiting and nausea for a day or so
130–170	25% chance of vomiting, other symptoms
180–220	50% chance of vomiting, other symptoms
270–330	20% deaths in 2–6 weeks, or 3 months recovery
400–500	50% deaths in 1 month, or 6 months recovery
550–750	Nausea within 4 hours, few survivors
1000	Nausea in 1–2 hours, no survivors
5000	Immediate incapacitation, death within a week

the 5,000 rad dose and spread it over a seventy-year lifetime, it may have little immediate effect, except to increase your cancer risk a bit. Depending on your lifestyle, or genetic heritage and predisposition, you may be more likely to die of some other factor rather than your cancer-induced radiation exposure. Astronauts, for example, are limited to 400 rads accumulated over their entire careers. If they absorbed this in one day, they would become extremely ill and have a good chance of dying from it.

To find actual instances of these kinds of high-level radiation dosages in humans, you have to look at what has happened to survivors of nuclear warfare or nuclear power plant accidents. In Hiroshima and Nagasaki, for instance, thousands of people were instantly vaporized as the radiation they absorbed raised their body temperatures to thousands of degrees in an instant. Many more people eventually died from the less than incandescent exposures they received. Even so, long-term studies of the survivors of the instantaneous 10–50 rem Hiroshima and Nagasaki dosages show that they actually have *lower* rates for leukemia and genetic defects in their offspring than the unaffected Japanese populations in neighboring cities. No one knows why.

Still, table 8.1 tells the average person very little about what they might expect from daily activities. To get this information, you have to look, for example, at the environmental dosages that have been tabulated by the International Atomic Energy Agency in Vienna, Austria. As you can see in table 8.2, the results are rather surprising. Compared to the biologically severe dosages in table 8.1, typical annual dosages are thousands of times smaller, and we have to employ a unit of 0.001 rem (one mil-

TABLE 8.2 Annual Radiation Dosages

Radon gas	130 millirem
Earth crust	60 millirem
Nuclear reprocessing plants	40 millirem
Cosmic rays	38 millirem
Medical	30 millirem
Food-water	23 millirem
Nuclear power industry	0.8 millirem
Fallout	0.1 millirem
Smoke detectors	0.05 millirem

NOTE: To compare this table with table 8.1, electromagnetic radiation dosages produce 1 rem of tissue dosage for each 1 of radiation dosage. For energetic particles, 1 rad of dosage can produce 10 rems or more of tissue dosage.

lirem) as a more convenient scale to gauge them (A chest X ray, on this scale, is about 20 millirems). In the international "SI" units, 0.001 rem equals 0.1 seiverts.

Topping the list is radon gas, a natural by-product of certain radioactive elements found everywhere in the crust of the Earth, especially in granite-rich rocks, and clays. You have probably never heard of radon gas until you bought, or sold, your first house. The radon gas hazard is the highest one we have to deal with, which is why basement radon gas monitors are a mandated part of home sales and purchases in the United States. This is a real and serious problem, not just another piece of legislation that the federal government wants to burden us with to make life complicated. The Environmental Protection Agency recommends that action be taken if radon levels exceed about 750 millirems per year. This usually means doing nothing more than installing a basement ventilation system to expel the stagnant, radiation-laced gases, which have seeped into the basement from the ground below the house foundation. And there we uncover yet another source of trouble.

The ground around your feet, the cement and brick in your homes also emit radiation from their infinitesimal loads of trapped radioactive debris to the tune of about 60 millirem per year, but this changes quite a bit depending on where you live. For example, in states like Georgia, California, Florida, and Maryland the terrestrial background radiation level is between 50–70 millirems per year, in Louisiana it is as low as 30, and in Colorado and South Dakota it can be as high as 115. The difference between living in Louisiana and Colorado is equal to an additional four chest X rays per year added to your lifetime total.

If you really want to live on the edge, you have to visit places like Kerala, India, where the thorium-rich sands give you a dose of 380 millirem every year, and in Guarapari, Brazil, where you get a sizzling 600 millirem per year. In comparison to the natural background sources and their variations, one wonders why so many people worry about one versus two extra chest X rays per year. If you want a big savings in exposure, just move to a seacoast town and forget about prolonged vacations in Denver, Brazil, or India.

From table 8.2 we can see some other surprising natural sources of radiation too. Just about every atom in nature has one or more radioactive variants called an isotope. When you add up the inhaled and ingested isotopes found in potassium and carbon, this alone is equal to 23 millirems per year. Your own body is itself a low-grade source of nuclear radiation. If you are worried about your radiation risks, you should probably stop eating bananas (rich in radioactive potassium isotopes). You should also give up smoking (40 millirem per year for a one-pack-a-day habit).

If these were the only natural sources of radiation, you would already have a typical annual exposure of nearly 250 millirems, or about ten chest X rays per year. There is hardly anything you can do about this except perhaps to ventilate your basement and change your eating habits. But, even so, there is another form of environmental radiation that you can do even less about.

One of the most unexpected sources of natural radiation doesn't come from the Earth at all. Instead, it rains down on our heads from the rest of the universe. Throughout the universe, massive stars grow old, die, and explode as supernova. These interstellar detonations fill space with particles that get accelerated to very high speeds and energies. Dense cores of imploded stellar matter–pulsars–are powerful magnetic accelerators that push particles' speeds to nearly that of light, hurling them deeply into the void. Even distant galaxies can have powerful magnetic fields that accelerate expelled stellar gases to very high energies. Over the course of billions of years, all of these sources suffuse space with a dilute, but energetic, gas of stripped atoms, electrons, and protons, all rushing about nearly at the speed of light.

As these particles stream into our solar system, the solar wind and magnetic field serve as a weak umbrella to deflect the less energetic cosmic rays. As the remaining higher-energy cosmic rays penetrate deeper into the solar system, individual planetary magnetic fields deflect still more of them. Eventually, the most energetic cosmic rays make it all the way into the Earth's atmosphere where they collide with nitrogen and oxygen atoms to produce secondary "showers" of energetic particles. These particles travel all the way to the ground and immerse the biosphere in a steady rain of particles day in and day out. Along the way, they also create radioactive carbon-14, which we ingest by the trillions of atoms every day. What this means for you and me is that a person living in Denver, the "Mile High City," or in Laramie, Wyoming, basks in an annual cosmic ray dosage of 120–130 millirems per year, while someone living in a seacoast town would only receive about 35 millirems. Travelers to remote mountaintops don't have to worry about bringing lead underwear to protect themselves. But it is true that prolonged stays on mountain peaks higher than fourteen thousand feet brings with it more than just the exhilaration of the experience. With each passing day, the cosmic ray drizzle bathes you with an invisible and relentless shower of radiation.

We have all heard, since grade school, that radiation affects living systems by causing cell mutations. The particles strike particular locations in the DNA of a cell, causing the cell to malfunction or to survive and pass on a mutation to its progeny. These accumulated "defects" seem to happen at a steady rate over the course of millions of years, and paleobiologists use "DNA differences" like a molecular clock to determine when species became separate. The DNA in chimpanzee and human blood hemoglobin tells a hidden story: that about five million years have passed since these species shared about the same DNA. The steady rain of cosmic rays and other background radiation seems to be the very engine that drives evolution on this planet. Inasmuch as we are all fearful of radiation, evolution on this planet requires it as the invisible agent of change. But sometimes the mutations are not beneficial to an organism, or to the evolution of its species. When this happens you can get cancer.

Cancer risks are generally related to the total amount of lifetime radiation exposure. The studies of Hiroshima survivors, however, still show that there is much we have to learn about how radiation delivers its harmful impact. Very large dosages over a short period of time (minutes to hours) seem not to have quite the deleterious affect that, say, a small dosage delivered steadily over many years does. Over the years, the National Academy of Sciences has looked into this issue rather carefully to find a relationship between lifetime cancer risks and low-level radiation exposure. What they concluded was that you get up to one hundred cancers per one hundred thousand people for every 1,000 millirems of additional dosage per year above the natural background rate. This has been translated by the Occupation Safety and Health Agency (OSHA) into "acceptable" risks and dosage levels for different categories of individuals and occupations.

OSHA assumes that the relationship between dosage and cancer death rates is a simple arithmetic proportion. If a dosage of 1,000 millirems extra radiation per year adds one hundred extra deaths per one hundred thousand, then as little as 1 extra millirem per year could cause cancer in one person per million. Although it's just a statistical estimate, if you happen to be that "one person" you will be understandably upset. No scientific study, by the way, has shown that radiation has such a "linear" impact at all levels below 100 millirem. The Hiroshima study, meanwhile, shows that a linear model may not apply to very large doses above tens of rems total body dose, but that's what the blind application of arithmetic shows. It's just an educated guess, but it has caused lots of spirited debates and probably a fair measure of anxiety, too; hence the common worry about what that annual chest X ray might do to you. I would be a lot more worried about that full-mouth dental X ray that can deliver from 500 to 900 millirems just so a dentist can fit you for braces or pull your wisdom teeth. According to the linear model of dosages, that lead blanket they like to put on you in the dentist's office probably does little to protect you from tongue or throat cancer.

The OSHA has worked out dosages for many different professions by balancing future cancer risks within the particular population in question against lifetime career exposures. For example, people who work with radiation, such as dentists, nuclear medicine technologists, or nuclear power plant operators, are given a maximum permissible dose limit of 500 millirems per year above the prevailing natural background rate. For those of us who do ordinary work in the office, factory, or store, the acceptable maximum dose is 1,000 millirems per year. As a comparison, if you lived within twenty miles of the Chernobyl nuclear power plant at the time of its 1986 meltdown, your annual dose would have been about 1,500 millirems per year during the first year, declining slowly as the radioactive isotopes in the environment decayed away. Some careers are worse than others for producing large incremental dosages compared to the environmental ones experienced at ground level. Surprisingly, one of those careers is that of airline flight attendant.

Jet airliners fly at altitudes above thirty-five thousand feet, which is certainly not enough to get them into space but more than enough to subject the pilots and stewardesses to some respectable doses when looked at over the course of their careers, and thousands of flights. A trip on a jet plane is often taken in a partylike atmosphere with passengers confident that, barring any unexpected accidents and food problems, they will return to Earth safely and with no lasting physical effects. But depending on what the Sun is doing, a solar storm can produce enough radiation to equal a significant fraction of a chest X ray's dosage, even at typical passenger altitudes of thirty-five thousand feet. Airline pilots and flight attendants can spend over nine hundred hours in the air every year, which makes them a very big target for cosmic rays and anything else our Sun feels like adding to this mix. According to a report by the Department of Transportation, the highest dosages occur on international flights that pass close to the poles where the Earth's magnetic field concentrates the particles responsible for the dosages.

Although the dosage you receive on a single such flight per year is very small—about one millirem per hour—frequent fliers who accumulate over one hundred thousand miles per year would also accumulate nearly 500 millirems each year. Airline crews who spend nine hundred hours in the air would absorb even higher doses, especially on polar routes. For this population, it is estimated that their lifetime cancer rate would be twenty-three cancers per one hundred people. By comparison, the typical cancer rate for ground dwellers is about twenty-two cancers per one hundred. But the impact does not end with the airline crew. The federally recommended limit for pregnant women is 500 millirems per year. Even at these levels, about four extra cases of mental retardation would appear on average per one hundred thousand women stewardesses if they are exposed between weeks eight to fifteen in the gestation cycle. This is a time when few women realize they are pregnant and when critical stages in neural system formation occur in the fetus.

Matthew H. Finucane, air safety and health director of the Association of Flight Attendants in Washington, D.C., has claimed that these exposure rates are alarming and demands that the FAA to do something about it. One solution is to monitor the cabin radiation exposure and establish OSHA guidelines for it. If possible, he also wants to set up a system to warn crews of unusually intense bursts of cosmic radiation, or solar storm activity during a flight. Meanwhile, the European Aviation Agency has gone even further, especially for crews using the high-flying Concorde.

Beginning in 2000, they now issue standard dosimetry badges to all airline personnel so that their annual exposures can be rigorously monitored. This is a very provocative step to take, because it could have a rather chilling effect on airline passengers. It might also raise questions at the ticket counter that have never been dealt with before: "Excuse me, can you give me a flight from Miami to Stockholm that will give me less than one chest X ray extra dosage?" How will the traveler process this new information, given our general nervousness over simple diagnostic X rays?

Consider this: during September 29, 1989, a powerful proton event on the Sun caused passengers on high-flying Concorde airliners to receive dosages of energetic particles equal to two chest X rays per hour. At the end of the flight, each passenger had silently received hundreds of millirems added to their regular background doses. Still, these occasional dosages the average person receives while flying, compared to the dosages we might accumulate once we land at another geographic location, are rather inconsequential over a lifetime. Compared to the quality of life that we gain in exchange for the minor radiation exposure we risk, most people will grudgingly admit the transaction is a bargain. Stat-

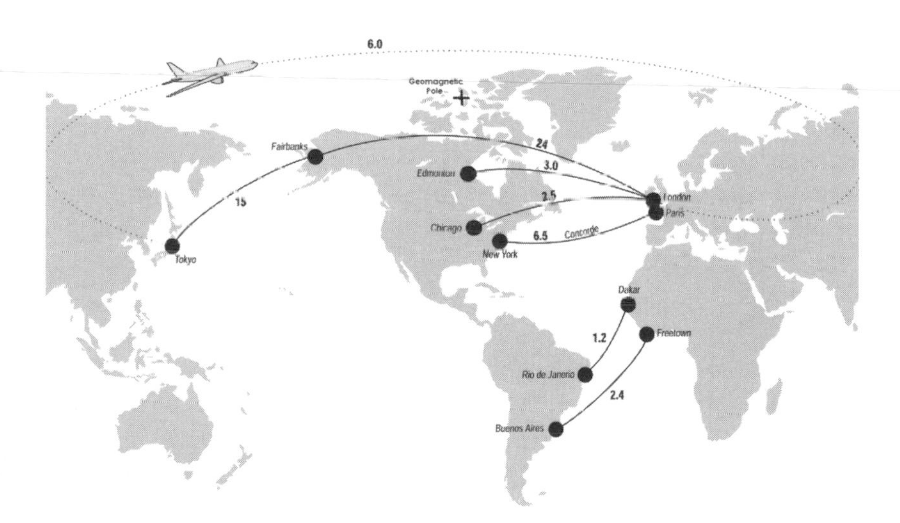

FIGURE 8.1 Radiation dosages for commercial airline flights flying at 35,000 feet; dosages (in millirem) due to cosmic ray given for one-way trips. Typical rates are 0.5 to 2 millirem/hour. Major solar flares (e.g., February 23, 1956) can produce total dosages as high as 1,000 millirem. *European Space Agency*

isticians who work with insurance companies often think in terms of the number of days lost to your life expectancy from a variety of causes. On this scale, smoking twenty cigarettes a day costs you 2,200 days, being overweight by 15 percent costs you 730 days, and an additional 300 millirem per year over the natural background dose reduces your life expectancy by fifteen days. There is, however, one human activity that seems to walk a precariously thin line between hazard and benefit.

Airline travel is far less of a hazard than space travel. Astronauts currently make routine trips to low Earth orbit in the Space Shuttle. Eventually, they may even take a few trips to Mars in the twenty-first century. Since the dawn of the Space Age, we have known that this environment presents a severe hazard for human health. Battered satellites bear mute testimony to the ravages of the various forms of radiation that penetrate their skins and do internal damage. Astronauts are given full briefings about radiation health risks before they start their journeys . . . and yet they rise to the challenge. On the other hand, the general public hears quite a bit about the medical health risks of space travel but, curiously, these risks are couched almost exclusively in terms of loss of bone density and peculiar cardiovascular changes. We never seem to hear much open discussion about astronaut radiation health effects. Compared to the tremendous intolerance we have on the ground for far less severe dosages, what astronauts are required to endure is positively horrific.

In space, radiation comes in three invisible packages delivered to the astronaut's doorstep. The worst of these are solar flares. At the present time, solar flares are completely unpredictable. By the time telescopes spot their telltale signs on the solar surface, millions of miles away, their deadly cargoes of X rays have already reached Earth orbit and caused a secondary shower of particles to flow in the skin of a spacecraft directed into the living quarters. A half-hour later, a burst of energetic particles begins to arrive. Both these components subject astronauts to high dosages of radiation and, depending on the amount of shielding, can pose a significant health hazard. During the Apollo program, there were several near misses between the astronauts on the surface of the Moon and deadly solar storm events. The Apollo 12 astronauts walked on the Moon only a few short weeks after a major solar flare would have bathed the astronauts in a 50–100 rem blast of radiation. This radiation level inside a spacesuit on the lunar surface would have been enough to make them feel ill several days later. But these are only the warm-up pitches in the celestial game of chance. Once every ten years or so, the Sun lashes out with even more powerful pyrotechnics, and we never see them coming.

The instant death scenario that dramatically unfolded in Michener's book was, perhaps, stretching the facts a bit too dramatically, but no space physicist finds fault with the basic idea that the most powerful solar flares are capable of killing unshielded astronauts. At issue is how long it might take, not the inevitability of the outcome. How often do we have to be worried about these super flares? Historical data on solar flare intensities provides some good clues.

Robert Reedy, a physicist at the Los Alamos National Laboratory, has spent much of his professional life wondering about this very issue, and his conclusions are comforting. Satellites, such as those in the GEOS and IMP series, have kept a close watch on the high-energy protons emitted by solar flares for decades. You can also find fossil traces of "solar proton flares" in the excess radioactive isotopes they produce in lunar rocks and terrestrial tree rings. What this far-flung data tells us is that flares in the same league as the August 1972 event happen only about once every ten years, usually just after the peak of a solar cycle. The long-term data also shows that solar flares ten times stronger than the August 1972 event have not been recorded in at least the last seven thousand years. James Michener's scenario of an instantly fatal flare may be rare, but biologically significant ones do happen rather often during a solar cycle. Given enough opportunity, and someone looking the wrong way at the wrong time, they are more than potent enough to cause severe radiation poisoning in an unshielded astronaut should their paths happen to cross in space and time.

When you look at the recorded solar flares since the late 1950s, it is easy to see some interesting trends in the numbers, especially when the information is presented pictorially. The calmest times for flares are within two years of sunspot minimum. It is as though even the Sun needs to rest from its labors, to shore up energy for the next round of activity. Sunspot maximum, with its tangled magnetic fields concentrated in numerous sunspots, seems to be the best season to go hunting for flares. Within two years of sunspot maximum, you have the greatest likelihood of having a medically significant flare within any given week. Near maximum, the typical time between significant (10 rem) flares can be about a month or so. The really major flares that deliver more than 100 rads to a space-suited astronaut happen once every year. But, like all flares,

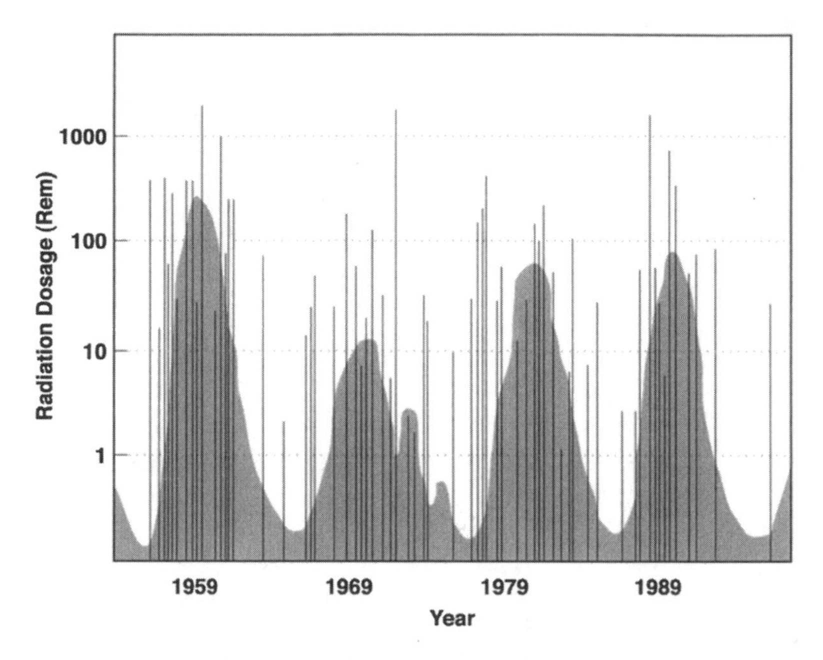

FIGURE 8.2 Energetic solar flares detected since 1958 and their estimated dosages for spacesuited astronauts.

they happen randomly, and no one really knows how to predict how powerful one will be before it reaches the Earth. The major flares that eventually kill you if you are unshielded happen every ten years on average. Solar Cycle 19, between 1955 and 1963, was a particularly nasty one, with no fewer than three flares that could have had some hazardous health effects. These happened during the years just past the sunspot maximum year. Cycles 20–22 were very similar in their flair statistics, but not as productive as Solar Cycle 19, which had the highest sunspot number at its peak. Apparently, the more sunspots a cycle has, the more opportunities there are for spawning potentially lethal or, at the very least, medically hazardous flares, during the declining years of the cycle.

In addition to solar flares, cosmic rays also pose a greater hazard in space than they do on the ground. The Earth's atmosphere is a natural shield against most of this radiation, to the tune of a four-yard thick slab of aluminum. You would hardly think that something as insubstantial as air could shield you from cosmic rays, but there is simply so much of it over your heads that it literally "all adds up." A Space Shuttle aluminum bulkhead, meanwhile, provides about two hundred times less shielding than this, but still enough to substantially reduce the health risk even from a flare that might be lethal outside the shuttle during a spacewalk.

Cosmic rays follow their own patterns of arrival here at the Earth; as a population and over the long term they are far more predictable than solar flares. The number of cosmic ray particles entering the Earth's environment does not remain the same but rises and falls exactly out of step with the solar activity cycle. When the Sun is very active near the peak of the sunspot cycle, its magnetic field is strong and penetrates farther out into the solar system, shielding the inner planets from some of the cosmic rays. When the Sun is less active during sunspot minimum, the solar magnetic field is drawn further in and cosmic rays can again penetrate into the Earth's environment.

Cosmic rays come and go with the solar cycle, and they also cause atmospheric carbon atoms to be converted into their radioactive form, called carbon-14. This is ingested into trees and other elements of the biosphere, so that traces of the rise and fall of the solar cycle are literally imprinted into the biosphere at the atomic level. Each of us bears a signature in our bodies of the solar cycle, encoded in the levels of carbon-14 we have ingested over our lifetimes. When very old trees are studied, we can actually use the carbon-14 in tree rings to reconstruct the sunspot cycle, thousands of years before the advent of the telescope.

Cosmic rays are a constant source of trouble for astronauts and spacecraft electronics, but the particles that flow in and out of the geospace environment are an especially bothersome population. The Earth's magnetic field traps high-energy particles in temporary belts or generates currents of particles like a magnetic dynamo. Closest to the Earth is a region called the plasmasphere, bounded by the most intense equatorial magnetic field lines. Within this moat of particles, high-energy electrons and protons in the Van Allen radiation belts flow along the magnetic field lines. They actually bounce back and forth along their northern and southern loops. At the same time, the electrons in these belts flow eastward while the protons flow westward in two great intermingled "ring currents." This region is instantly lethal and would zap an astronaut with 1,000 rems per hour if unshielded. Beyond the Van Allen belts and the plasmasphere, the rest of geospace environment contains a shifting patina of particles and fields that adopt part of their populations from the impinging solar wind that constantly streams by just beyond the magnetopause boundary.

The flows of these particles are exquisitely complex and far from random. Particles from Earth's own atmosphere are levitated out of the ionosphere in great polar fountains and are deposited in the plasmasphere. They buzz about like a superheated fog of matter held at temperatures of thousands of degrees. For many years it was thought that the solar wind supplies the van Allen belts with their particles, but satellite measurements soon showed that the chemistry was all wrong. The solar wind contains mostly hydrogen and helium nuclei, not oxygen. Instead, the ultimate source for the van Allen belt particles seems to be the Earth itself. Through a series of steps that are still not understood, these atmospheric particles are accelerated to very high energies. It is somewhere in these murky processes that they become transmuted into hazards for living organisms, but only if you venture into their lair.

So, with all these populations of particles ready to penetrate astronauts and cause them harm, you would think that very stringent health restrictions would be placed on astronauts as they leave the protective layers of the atmosphere. For a variety of technical reasons, OSHA pegs the career annual dosages at a far higher rate for astronauts than for the average person, or even the much maligned nuclear plant worker. Their exposures to solar flares, cosmic rays, and trapped particles are confined to only a few weeks at present. Besides, the risk is seen as going with the territory. Career dosage limits are set at an astonishing 100–600 rem depending on the astronaut's age and sex, but at no time can the doses exceed 50 rem per year, or 25 rem during any thirty-day period.

As enormous as these limits may seem to us ground dwellers, they are probably a rather generous lifetime limit for now–especially considering that typical mission-accumulated dosages have rarely exceed eight rem, as table 8.3 shows.

The total radiation dosage that an astronaut receives depends on a number of factors that are different from mission to mission. Being closer to the Earth (Gemini versus Space Shuttle) allows greater protection by the magnetic field and atmosphere of the Earth and keeps you farther away from the inner edge of the van Allen radiation belts. A week-long junket to the Moon exposes you to far more cosmic ray and high-energy particle damage than LEO. Most of this is because you have to travel through the van Allen Belts themselves to get there, although the transit time through the belts takes less than an hour. Also, just staying in space a long time, no matter where you are, is also a major factor for increasing radiation dosage, as can be seen in the data from the Skylab missions.

TABLE 8.3 Accumulated Manned-Spaceflight Radiation Dosages

Mission	Date	Altitude (miles)	Duration	Radiation dosage (millirems)
Mercury 9	5/15/63	166	34h	27
Gemini 3	3/23/65	140	4h	20
Gemini 4	6/3/65	182	97h	45
Gemini 5	8/21/65	219	190h	177
Gemini 6	12/4/65	185	320h	25
Gemini 7	12/15/65	167	25h	150
Gemini 8	3/16/66	169	10h	10
Gemini 9	6/3/66	185	72h	18
Gemini 10	6/18/66	165	70h	840
Gemini 11	9/12/66	175	71h	25
Gemini 12	11/11/66	168	94h	15
Apollo 7	10/11/68		260h	160
Apollo 8	12/21/68		147h	160
Apollo 9	3/3/69		241h	200
Apollo 10	5/18/69		192h	480
Apollo 11	7/16/69		195h	180
Apollo 12	11/14/69		10d	580
Apollo 13	4/11/70		6d	240
Apollo 14	1/31/71		9d	1140
Apollo 15	7/26/71		12d	300
Apollo 16	4/16/72		11d	510
Apollo 17	12/7/72		12d	550
Skylab-2	5/25/73	433	28d	1980
Skylab-3	7/28/73	433	60d	4710
Skylab-4	11/16/73	433	84d	7810
STS-1	4/12/81	269	2d	20
STS-2	11/12/81	254	2d	15
STS-3	3/22/82	280	8d	461
STS-41A	11/28/83	250	10d	141
STS-41C	4/6/84	498	7d	689
STS-51D	4/12/85	454	7d	472
STS-51J	10/3/85	509	4d	513
STS-29	3/13/89	317	5d	48

NOTE: 10 millirem is about equal to one chest X ray.

The MIR space station has been inhabited for over a decade, and according to Astronaut Shannon Lucid, the daily dosage of radiation is about equal to eight chest X rays (160 millirem) per day. Typical MIR crew rotations are about 180 days, so a mission dosage can be up to 30 rads. This is about in the same ballpark as estimates by Tracy Yang at the Johnson Space Flight Center. The constant radiation dosage that human bodies absorb causes chromosomal damage, and, for MIR cosmonauts, Yang discovered that this wear and tear implied dosages up to 15 rads. This is about equal to a thousand chest X rays over the course of the mission. Meanwhile, Ts. Dachev and his colleagues at the Space Research Institute in Bulgaria arrived at similar radiation exposure levels from actual dosage measurements on the MIR. Each traverse through the South Atlantic Anomaly provides 2 millirads behind the MIR bulkhead. Since there are about eighteen orbits per day in a 180-day shift, this works out to a total mission dosage of about 55 rads. The bottom line is that prospective International Space Station astronauts will probably receive somewhere from 15–50 rads of radiation per shift as they go about their work. Eventually, the laws of chance dictate that solar flares and human space activity must inevitably coincide with potentially haz-

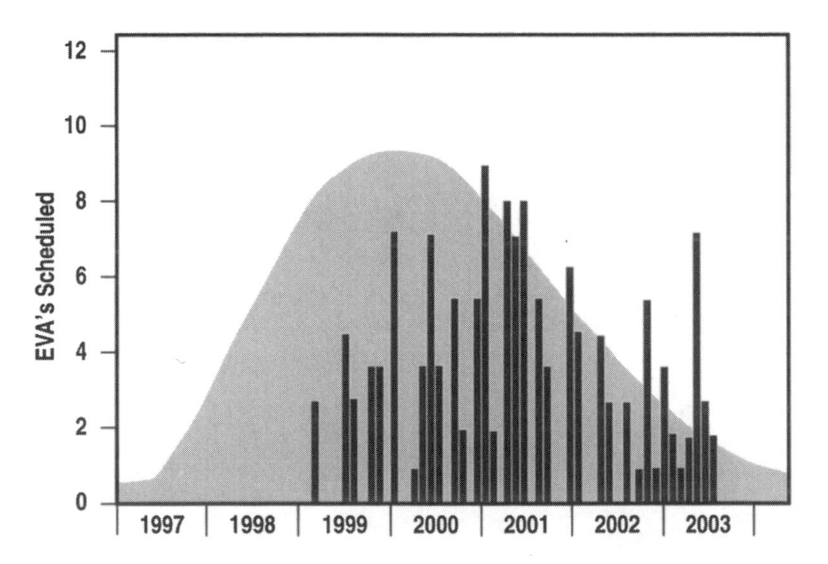

FIGURE 8.3 The schedule of spacewalks for the construction of the International Space Station during Cycle 23.

PLATE 1. Aurora from space taken by Space Shuttle (STS-39) crew in 1991. NASA

PLATE 2. Magnetic field lines near a sunspot taken with the NASA *TRACE* satellite. *TRACE*

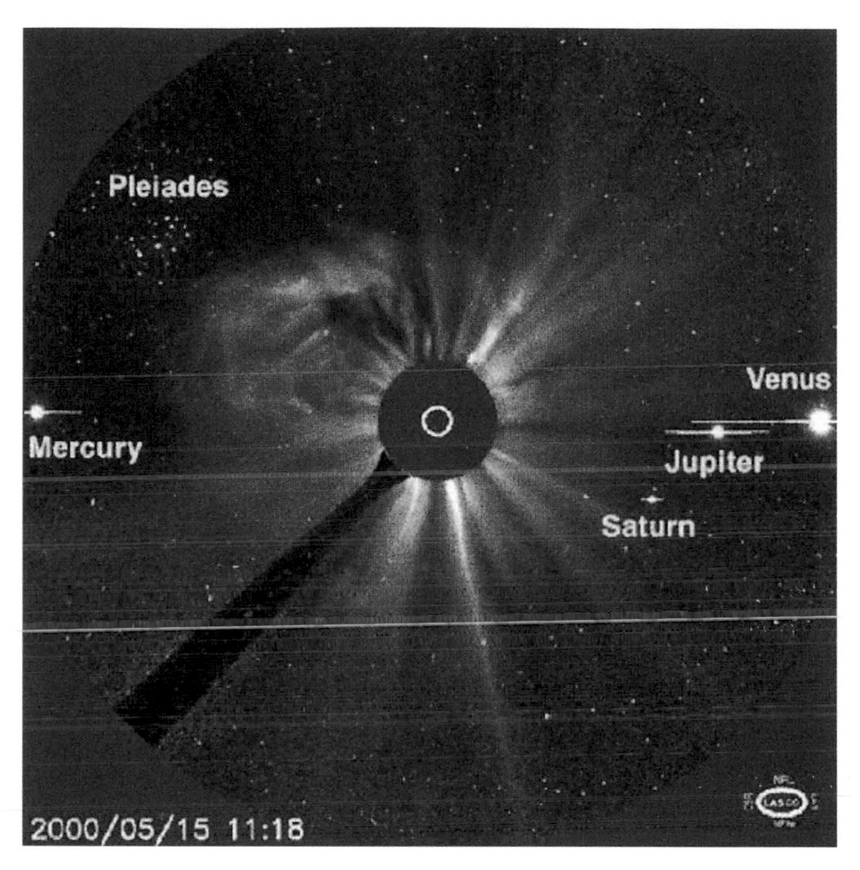

PLATE 3. A glimpse of the solar corona showing a cloud of ejected material. Also in the view are several planets near the Sun around the time of the May 5, 2000, "End of the World" conjunction, together with the Pleiades star cluster. This planetary conjunction was not visible from the Earth because of the closeness of the planets to the Sun. The NASA/ESA *SOHO* satellite's LASCO instrument was able to artificially "eclipse" the Sun to reveal the planets close to the solar limb.

PLATE 4. A transformer damaged by a geomagnetically induced current that literally "fried" the transformer's insulation. *John Kappenman, Metatech Corporation*

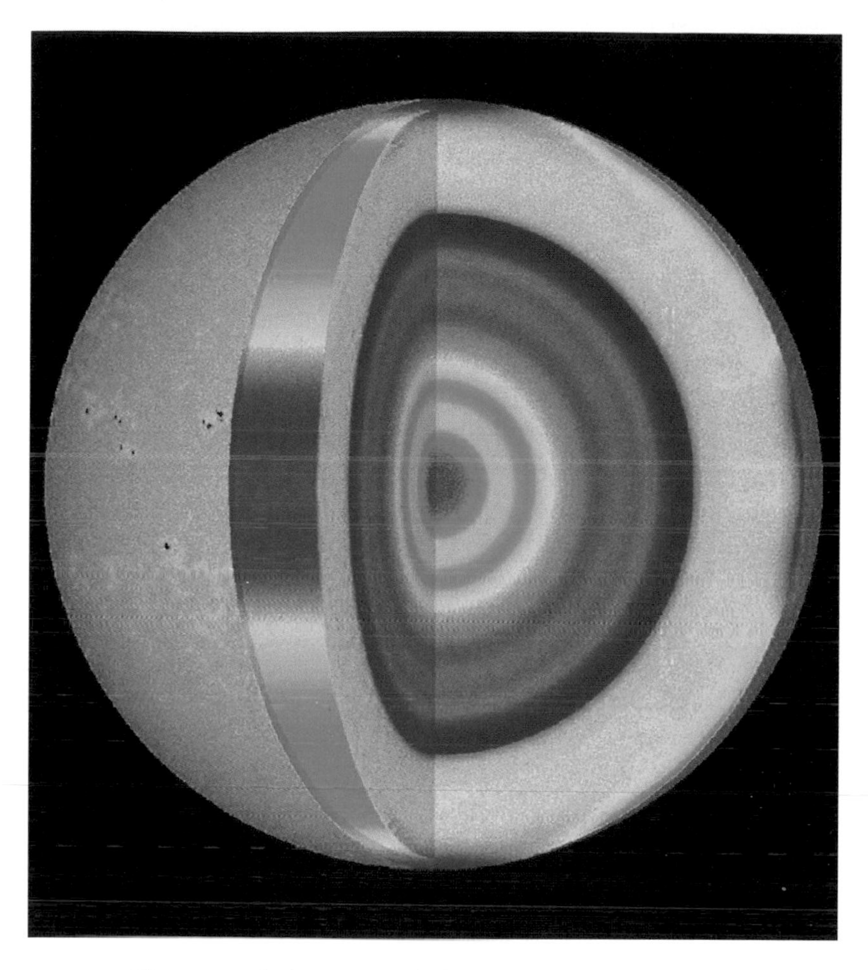

PLATE 5. The origin of solar activity is believed to occur deep inside the Sun. This computer model shows the interior of the Sun detected with the Stanford University helioseismometer on the *SOHO* spacecraft. Over one million points were measured on the surface. The red colors show where the computed speed of sound is higher than theoretical models predict and reveal complex interior layers. The Sun's core is detected at a temperature 0.1 percent lower than fifteen million degrees, suggesting a slightly reduced solar output. SOHO/*MIDI*

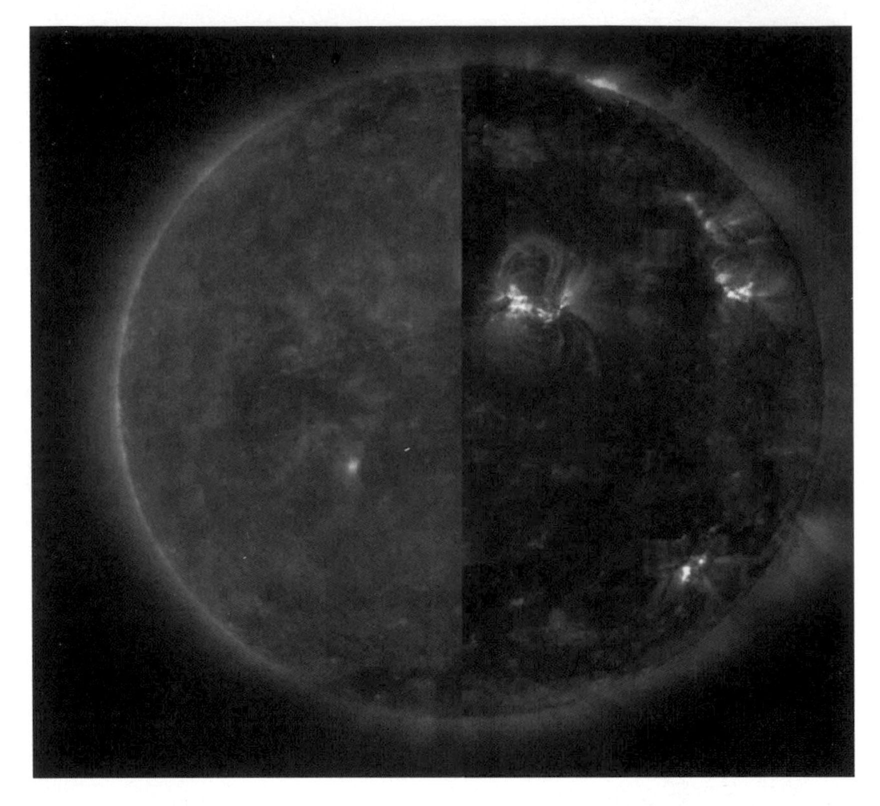

PLATE 6. The changing face of the Sun between sunspot minimum in 1996 (*left side*) and sunspot maximum in 1997 (*right side*), obtained with the *SOHO* EIT instrument. Note that the solar minimum surface is smoother and less structured by active regions compared to the Sun near solar maximum.

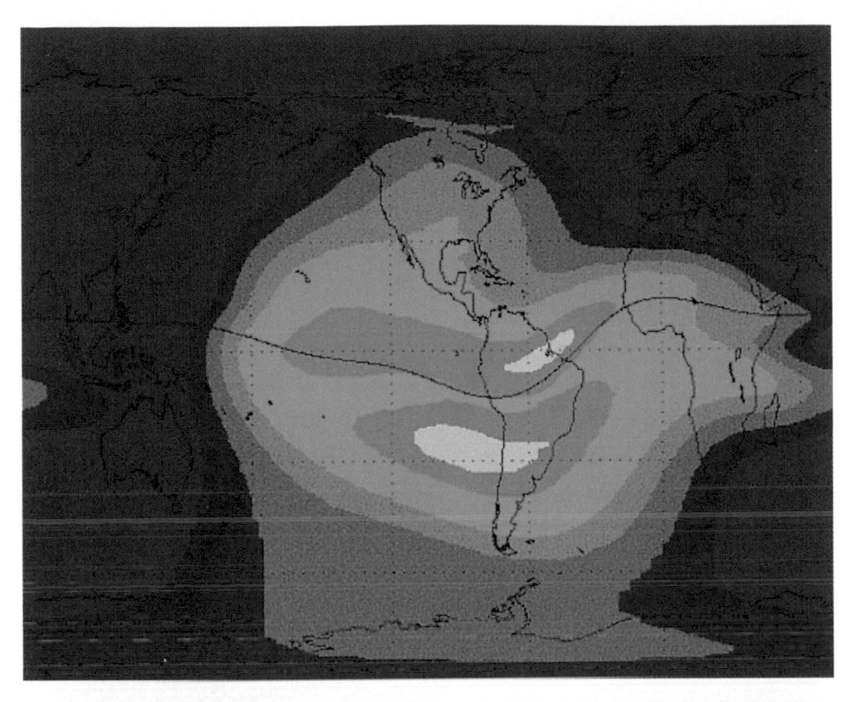

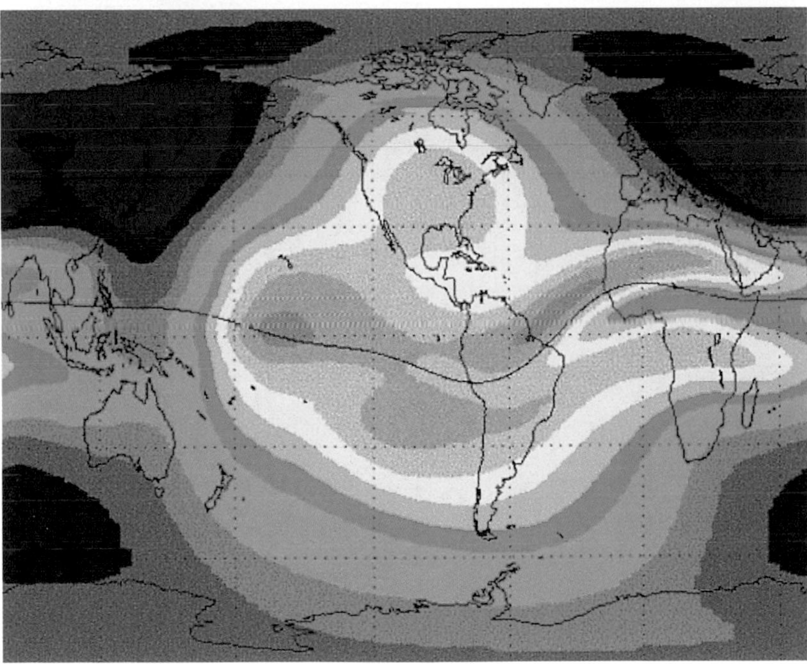

PLATE 7. Solar minimum and maximum reflected in the changing properties of the Earth's ionosphere. Ionospheric electron density maps for December 1995 near solar minimum (*top*) and December 1990 near solar maximum (*bottom*). *National Geophysical Data Center, NOAA*

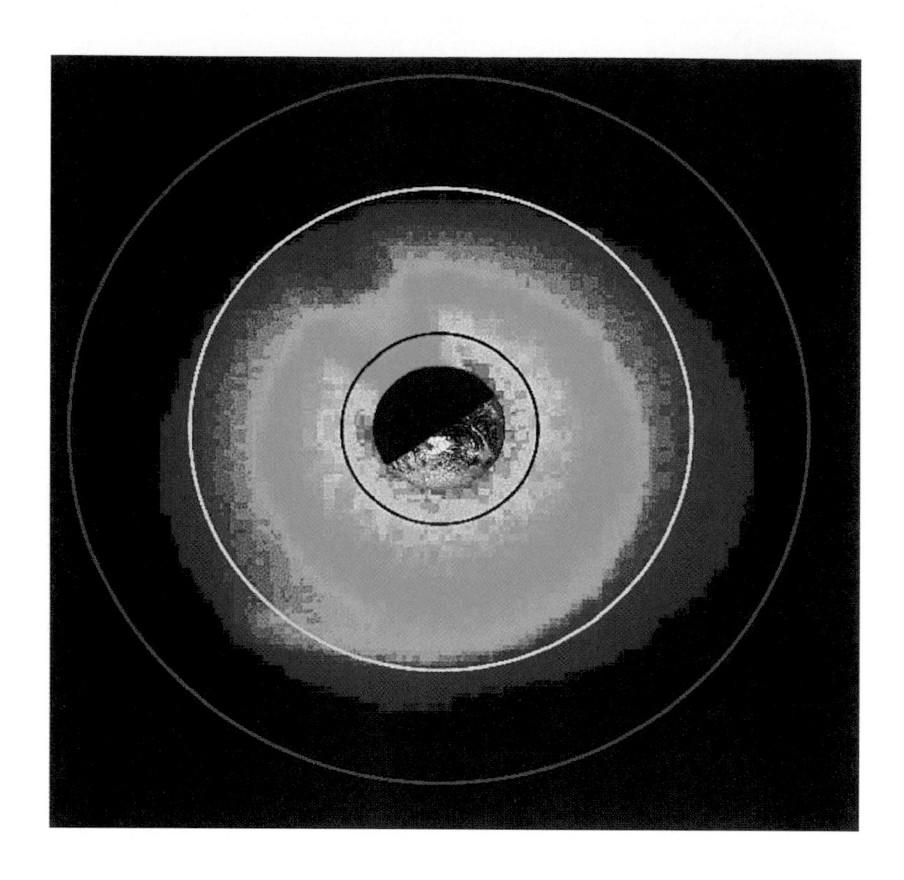

PLATE 8. The Earth's outer atmosphere extends thousands of miles into space and provides a constant friction to satellites. This image is a model produced by the NASA *IMAGE* satellite team to anticipate what the satellite's geocoronal imaging camera (GEO) will see. Also shown are the orbits of communication satellites (*red circle*), and typical distances to *GPS* satellites in MEO orbits (*yellow*), and *LEO* satellite (*black*) networks. IMAGE

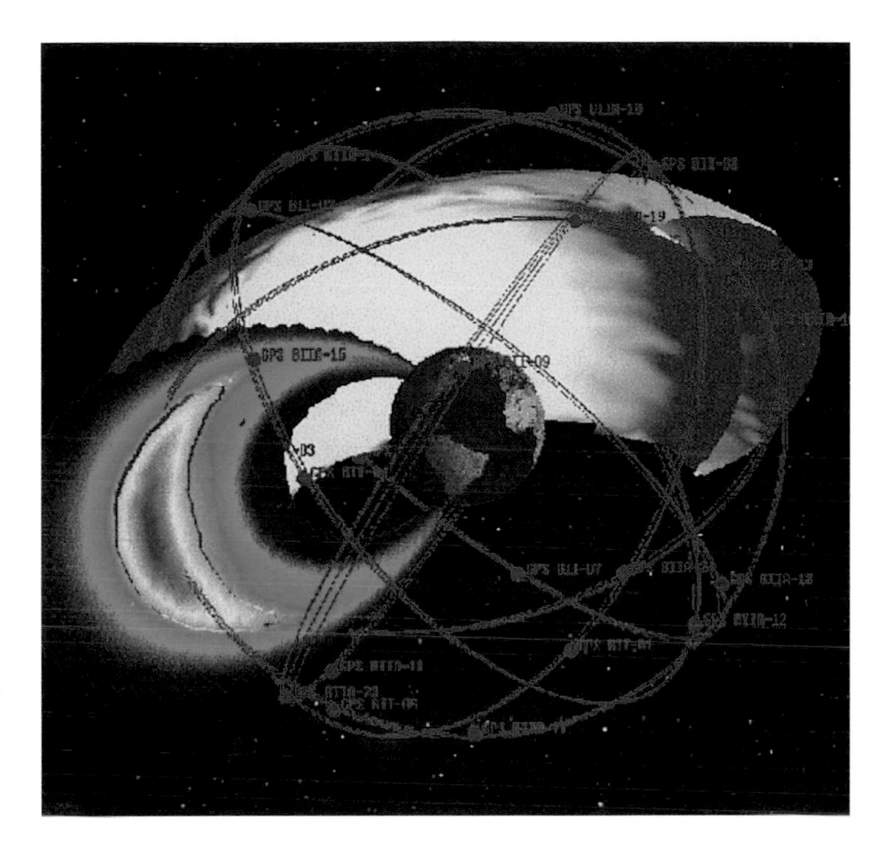

PLATE 9. The Global Positioning System satellites actually orbit inside the donut-shaped inner Van Allen radiation belt, seen here in cross-section. This model shows the location of the high-energy proton belt. Red colors indicate the most intense concentrations of particles. *U.S. Air Force*

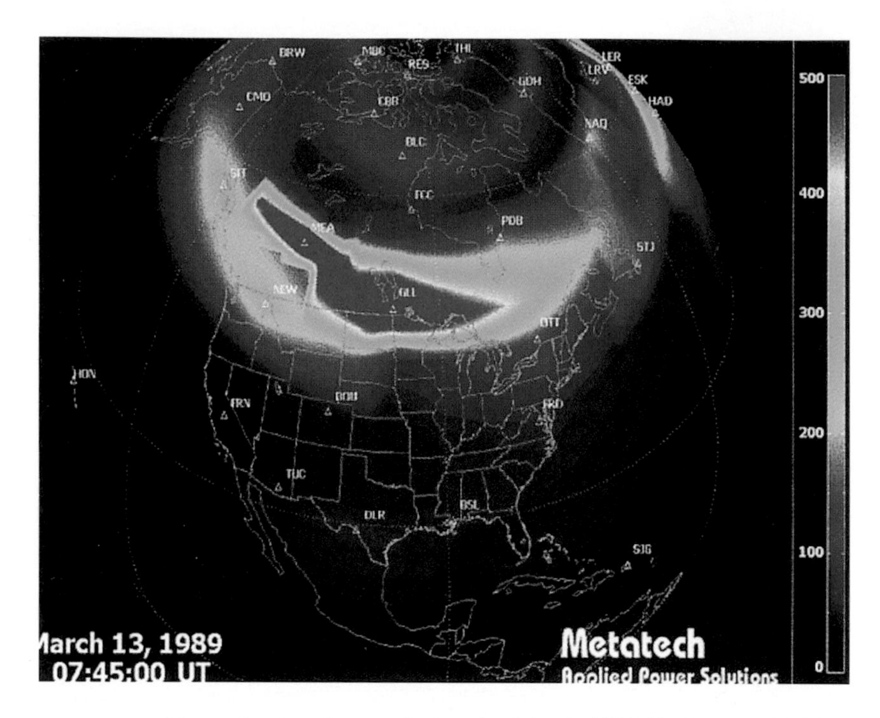

PLATE 10. Magnetic disturbances during the March 1989 blackout at 07:45:00 universal time. Within seconds, the disturbance traveled eight thousand miles from Canada to Great Britain. *John Kappenman, Metatech Corporation*

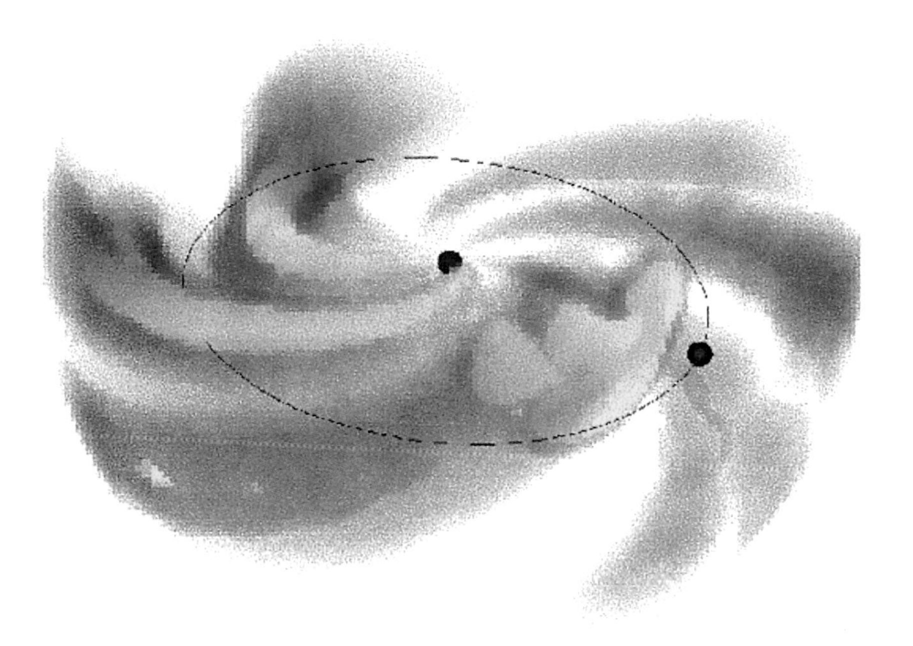

PLATE 11. Images of the solar wind detected by scattered sunlight. Note the pin-wheel-like structure of material ejected from coronal holes and other active regions on the solar surface. The Earth's orbit is drawn to scale. Earth and Sun dots are not drawn to scale. *Bernard Jackson, University of California Santa Barbara*

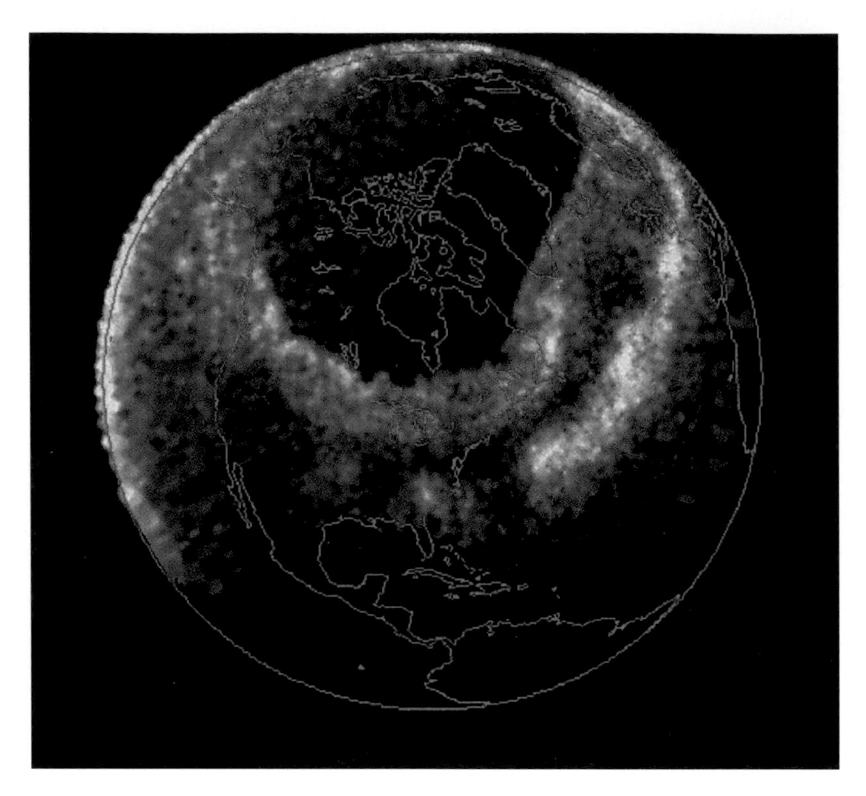

PLATE 12. From space, the aurora borealis appears as an oval of light centered on the north magnetic pole. This is a *Dynamics Explorer* satellite view of the auroral oval during the March 1989 Great Aurora. Note the large equatorward extent into North America and the especially bright knot of auroral emission over South Carolina. *Dynamics Explorer Satellite*

ardous consequences. NASA launch schedules are not known to take space weather conditions into consideration.

During the April 12, 1981, Great Aurora, *STS-1*, commanded by Robert Crippen on its maiden flight, was launched while the storm was actually still in progress. Astronauts were told by NASA that the radiation levels inside the Shuttle might be high enough to trip the smoke alarms, although this never actually happened. The actual dose accumulated by *STS-1* astronauts was rather small compared to other flights because they only spent two days in orbit. Eight years later, during the October 1989 storm, Space Shuttle *Atlantis* astronauts experienced light flashes in their eyes during the storm events, and they retreated to the interior of the Shuttle. This did little good, and the light flashes were still seen, accompanied by eye irritation as well, especially during the episodes of high radiation fluxes. These light flashes are charged particles passing through the Shuttle bulkhead and through the eyes of the astronauts, causing luminous streaks. At about the same time, solar storms towards the end of 1989 caused MIR cosmonauts to accumulate in a few hours a full year's dosage limit (probably exceeding 25 rads).

So far, we have been discussing astronauts working and living inside air-conditioned spacecraft in shirtsleeves. Normally, astronauts and cosmonauts do spend the vast majority of their time inside the shielded spacecraft with very few space walks. Space walks are still considered the most risky thing that an astronaut can be called upon to do. Little wonder, when you consider what kinds of hazards can be lurking outside the hatch. But soon these expectations will, at least temporarily, be a thing of the past as we exit the peak of Solar Cycle 23. The Space Station will be assembled in LEO orbit at an altitude of 220 miles, and its assembly will involve a projected 960 hours of space walks by eighteen astronauts. There will be about one hundred space walks planned during thirty-nine assembly flights between 1999 and 2003.

The vulnerability of the astronauts to solar flares is a major concern by EVA planners because they can occur with little warning. Conceivably, the very tight assembly and EVA schedules for the ISS may slip by weeks or months or more if the Sun decides to favor us with a potentially hazardous state during the missions. The actual probability that an astronaut will be affected by a solar flare large enough to be medically important is rather low, but it is not zero. The smaller, more frequent flares, which NOAA's Space Weather Center classifies as "S3," happen about once a year. Astronauts would have to stay inside the Space Shuttle for several days while the radiations subside. For the more powerful flares in classes S4 and S5, which happen up to once every three years, the mission may be aborted altogether. Caught outside with a once-per-cycle S5 flare, an astronaut could find him or herself removed from further space duty . . . or from the world of the living.

Radiation exposure problems will, of course, not end with the assembly of the ISS. Once completed, the ISS will be occupied by up to eight astronauts in shifts lasting about five months each. A five-month stay, at a typical dose rate experienced by the MIR cosmonauts, leads to an accumulated dose of up to 25 rads per shift. This is comfortably below the 400-rad lifetime limit set by OSHA and the 50 rad limit for annual dosages. But a single solar flare could, as we have seen, change this in a hurry. For the longer stays in space needed for interplanetary travel, measured in years, the exposure situation is much worse and far harder to anticipate.

Voyagers to Mars will find themselves utterly unprotected by the Earth's magnetic field, whose invisible cloak at least shielded them from some of the cosmic ray and solar particles. The shielding needed to reduce flare dosage levels below the OSHA astronaut health limits is substantial and can easily exceed many tens of tons. When you consider that current launch vehicle technology allows for shipping rates to Earth orbit between $5,000 and $15,000 per pound, shielding weight is bought at a premium. In the end, the Mars crew will probably still receive between 100–300 rems of accumulated dosage during the 500–1000-day Mars mission, depending on when they started their journey and the level of solar storminess they experienced.

With all these potential risks to technology and health to worry about, just how bad have things become during the current cycle of solar activity, Solar Cycle 23?

9 Cycle 23

During 1999 and 2000, we really expect some wild rides. We really don't know what effects we are going to see.

—JoAnn Joselyn, Cycle 23 Project, 1996

The instruments on board NASA's Solar and Heliospheric Observatory (*SOHO*) were routinely keeping watch on the Sun on April 7, 1997, when the Extreme Ultraviolet Imaging Telescope (EIT) camera picked up a typical garden-variety, class-C6 solar flare in progress. Scientists back on Earth watched while a shock wave from the flare passed through the local gases in the solar corona like the waves from a pebble dropped into a pond. It was a beautiful event to watch, looking for all the world like some artful animation rather than the awesome detonation that it actually was. In minutes, a ring of compressed gases had spread to engulf a patch of the Sun as big as the Earth. Radiation sensors onboard the geosynchronous *GOES* weather satellites detected a rain of flare particles minutes later; meanwhile, radio telescopes began to detect the telltale radio waves from a Type II burst on the Sun. The CME, in its haste to leave the Sun, had shocked and compressed solar plasma ahead of it, snowplowing them into walls of stripped atoms and magnetic fields that emitted powerful blasts of radio waves. At 10:00 A.M. EDT, as the shock wave spent itself, the LASCO instrument witnessed a major CME grow to the size of the Sun and larger.

Three days later, on April 10, 7:00 P.M. EDT, the *WIND* and *SOHO* satellites, parked one million miles from the Earth toward the Sun, started to feel the direct impacts of energetic particles from the CME. The faint signals from the compressed interplanetary wind had already been perceived a few hours earlier. Ground-based magnetometer readings from CANOPUS, the Canadian magnetic observatory network, started to sense major changes in the Earth's field heralding a Large Storm Commencement at 10:00 P.M. EDT. Meanwhile, the POLAR satellite had already seen auroras begin to grow on the dayside of the Earth at 2:50 P.M. EDT. By 5:26 P.M., intense nighttime aurora could be seen in New Hampshire and Massachusetts as the aurora slid past the U.S.-Canada border and plunged into the Lower-48. Many amateur photographers reveled in spectacular opportunities to capture on film both the dazzling auroral curtains and the history-making comet Hale-Bopp.

The great series of domino events tracked by NASA satellites, literally from cradle to grave, prompted scientists to release a press announcement on April 8 that predicted the real meat of this CME would harmlessly pass about a few million miles below the plane of the Earth's orbit. At best, it would be a glancing blow and, most probably, not a direct hit. As seen from the Sun, hitting the Earth is not exactly a turkey shoot, even with a million-mile-wide bullet. The magnetosphere of the Earth extends over one hundred thousand miles from the center of the Earth and has about the same apparent size as a dime held at thirty feet. Even though CMEs are huge, the Earth is such a small target you really have to get CME and solar flares pointed right at the Earth before there is a good chance of any physical contact happening.

The news media were especially fascinated by this cosmic salvo. The spectacular satellite images of its genesis, millions of miles away, made the CME near miss almost irrelevant. It really didn't matter if the storm would only be a glancing blow this time. April 10 turned out to be a big news day for this cosmic nonevent, with nearly all the major national and international newspapers carrying some kind of story about it. Some reporters, unfortunately, rushed into press with rather sensational stories such as the *New York Times*, "Storm on Sun Is Viewed from Spacecraft: First Detailed Look at Solar Event That Could Effect Life on Earth," which was datelined April 9 and published on April 10, 1997. Meanwhile, CNN and Yahoo!News reassuringly reported in their on-line news services that "Solar Flare Small After All, Poses Little Damage" (CNN) and "Solar Storm's Full Force to Miss Earth" (Yahoo). Even the *Boston Globe* reported, "Not much flare to this solar event, experts say." NBC, CBS, and CNN News carried interviews with George Withbroe, chief of the NASA Office of Space Science, and Nicola Fox, a scientist at NASA's Goddard Space Flight Center who coordinates the Global Geospace Science program. Nearly every news report mentioned possible technology impacts should the CME actually hit the Earth, including electrical blackouts and satellite outages.

Our Sun traces a dependable path across our skies every day, yet only very recently have we documented that it has its share of stormy days. For the last three centuries, solar activity levels have come and gone in a roughly eleven-year pulse beat that we have actually grown to expect. Even our biosphere shows the unmistakable traces of these cycles resonating in everything from carbon-14 abundances in tree rings to global precipitation patterns and coral layering. Our eyes never see the Sun brighten or dim, nor are we even remotely aware that the Sun cycles back and forth from stormy to quiescent. It is a small cause that manages to have a big effect on the hidden aspects of our environment. The fulcrum lies somewhere in the dark spaces between the solar photosphere and our own murky comprehension of the Sun Earth connection.

Just knowing about the solar cycle has been a promising first step in figuring out what the Sun will do over the long haul, even though the average person on the street is hardly aware that these cycles happen. The solar cycle, however, is a poor barometer of what we should expect the Sun to do for us tomorrow, and that's what is most interesting to satellite owners, astronauts, and electrical utility managers. Within each regular cycle, the Sun is actually rather temperamental. It hurls flares and billion-ton clouds at us almost at random, dissipating, with each blast, any sense of predictability.

By the time you read this chapter, one thing is certain: you will be near the end of this book but only about halfway through the current sunspot cycle–Cycle 23. If you were a solar physicist, a satellite owner, or a general in the Armed Forces, the question that you would be asking by now is, "Just how *bad* will the rest of this cycle be?" The answer depends on what you are concerned about. If you are worried about your communications or espionage satellite: "Will a flare erupt in the next day or so, and cause a satellite anomaly?" If you are trying to plan for next year's budget: "Are we in for a bad solar 'summer,' with many more opportunities for weekly flares and technological difficulties?" Either way, it is hard to know with any certainty. If a flare as bad as the Cycle 20 *Apollo 17* near miss arrives during International Space Station construction, astronauts could be seriously affected. We could also be treated to a Quebec-style blackout as we were in Cycle 22, which could cost several billion dollars to recover from. Truly exceptional solar events seem to favor the declining part of the solar cycle, which for us will arrive between 2001 and 2006. So we do what became a reflex reaction to uncertainty in the second half of the twentieth century:

We set up a committee to study the problem.

For much of the past century, groups of scientists gathered together to try to guess how bad each cycle would be. With so many expensive undertakings on the agenda for this cycle, NASA and NOAA continued this long tradition in 1996 by setting up the Solar Cycle 23 Project. The first thing this panel did was contact the rest of the astronomical community and invite everyone to send in what they thought Cycle 23 might be like. The catch was that they also had to describe, in detail, what method they used to make the prediction. There was no reason for the panel to try and reinvent the wheel in solar activity forecasting when the community they represented had already worked this particular problem for decades. The panel's request brought to their table no fewer than twenty-eight separate methods–by some estimates, nearly as many methods as there are researchers in this particular field. Only a few of the methods, by the way, had anything to do with the popular image of counting sunspots. Some of them tracked the rise and fall of geomagnetic storms here on Earth, others followed the total amount of radio power from the Sun at a wavelength of 10.7 centimeters. Ultimately, as for an ancient traveler in Italy, all roads lead to Rome.

The biggest problem everyone had to face was that in 1996 the Sun was smack in the middle of the activity *minimum* between Cycles 22 and 23. Getting a reliable prediction for Cycle 23, without even a year's worth of data on the new cycle, wasn't going to be easy at all. Scientists are not at all new to this kind of a situation. They usually have to face frustration, and uncertainty, every day as they conduct their research. When you work with limited information, a common circumstance in space research, you often have to bridge the gaps by using past experience and the collective knowledge of physical science as a guide.

So the panel weighed the uncertainties in each of the twenty-eight methods and how well they had anticipated previous cycles. In the end, they were prepared to say that the time of the maximum for Cycle 23 would be around March 2000. They also hedged their bet by offering the alternative prospect that solar maximum could happen as late as June 2001 or even as early as the summer of 1999. The March 2000 prediction, however, was close to the average made by most of the methods, so that's the one the panel favored. The panel also predicted how high the activity levels would rise. Their prognosis was that Cycle 23 probably wouldn't be as bad as Cycle 19, but it might be at least as bad as Cycle 21, with its roughly 150 spots near maximum. Incidentally, the panel didn't consider the possibility that this "millennial" sunspot cycle might be the Sun's last one. This is not such a far-fetched possibility at all. Before 1700, telescopic observers of the Sun detected few sunspots. Their meticulous observations provided no hint to later generations of astronomers of any periodic rises and falls. There was a real chance that we could wake up in 1997 or 1998 with no new cycle anywhere in sight.

Once the new cycle began, the panel's predictions for the minimum, average, and maximum Cycle 23 activity curves soon had data starting to crawl up as though the Sun were navigating a three-lane highway. It was pretty obvious by June 2000 that the predicted trends were running a bit higher than expected. Monthly sunspot averages were just below the curve the panel had offered up as their minimum activity prediction. Rather than the high-speed lane, the Sun slowed to the breakdown lane to get through the turn of the millennium.

So far, the ascent up the jagged curve of Cycle 23 has been quite a wild ride. The years 1997 and 1998 had their share of spectacular solar eruptions and auroral displays. On the other hand, 1999 was uneventful for the Earth despite the fact that over one hundred CMEs were being ejected by the Sun each month. The good news was that Cycle 23 was not going to be a major storm period this time–at least not like the levels reached in Cycles 21 and 22. When you factor in the details of individual flares and CMEs, the dossier for Cycle 23 has turned out to be dramatically more complicated than what a simple count of sunspots would tell you. It is the day-to-day engagements with solar flare and CME "bullets" that can cause harm.

As scientists settled in to watch their data, eagerly anticipating new breakthroughs in solar research, the news media also developed an interest in keeping watch on the Sun. In fact, Cycle 23 has been hawkishly watched in a way that no other previous cycles have been. Armed with the latest spectacular imagery from NASA's satellites, it has been much easier to anticipate when bad things might be brewing, because you can actually *see* them happen days in advance. Even school children can visit NASA web sites to view images of today's solar surface and make their own predictions. Before the first spots of this cycle started to appear, *Time* magazine announced, in 1996, “Cosmic Storms Coming.” A year later, *Space News* also cautioned their readers that “U.S. Scientists Warn of Rise in Solar Flares.” NASA held a formal press conference at the Jet Propulsion Laboratory on February 15, 1996, which was heralded as a “Briefing Directly Linking Solar Storms to Disruptions on Earth.” The briefing topic was about seventy years too late, given all the impacts already endured during the twentieth century.

Cycle 23 officially began in September 1996 when the first spots of the new solar magnetic polarity cycle were detected, followed by a month when not a single new spot was seen. As the first spots of this cycle began to slowly appear at high solar latitudes, like soldiers nervously sitting in a foxhole waiting for the first rounds to fly, engineers and scientists braced themselves for the inevitable solar onslaught. They didn't have long to wait. Barely three months after the start of the new cycle, the first major solar event leaped out at the Earth and, by many accounts, claimed its

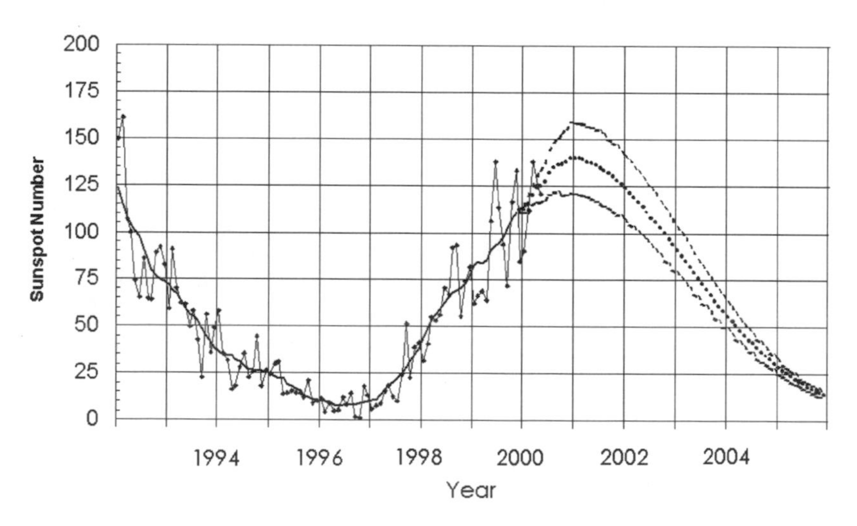

FIGURE 9.1 The progress of Cycle 23 by June 2000 in terms of the sunspot numbers. The jagged line shows the average monthly sunspot count. The smooth line shows a smoothed version of the sunspot number counts. The three dotted lines show the currently forecasted ranges of the remaining sunspot cycle. The lower prediction seems to be favored, and the most likely time of the sunspot maximum seems to be between November 2000 and March 2001.
SOURCE: *NOAA/SEC*

first technological casualty: one of the key satellites of our communication infrastructure—*Telstar 401*.

This was followed by the April 10 near miss of the Earth by another CME, which was highly publicized. NASA scientists, armed with their new satellite technology, had seen the January 1997 event coming, and with the April 1997 event they were now "two for two" in having predicted the course of major solar storm events during the new sunspot cycle.

A month after the April 10 near miss, *Space News*, a much read weekly newspaper of the space community, carried a short article about a major new-generation satellite that had encountered space weather difficulties at about the time this solar event reached Earth's orbit. The *Tempo-2* satellite, equipped with the latest in high-power, gallium-arsenide solar arrays, lost 15 percent of its operating power on April 11, and this was directly credited to the solar storm by Loral's space systems division, which manufactured the satellite. In a May 1 statement, company spokesperson David Benton announced, "We have evidence from sensors on the satellites that there was a space event in the vicinity of the Tempo satellite at the time of the disturbance."

The *Space News* article also mentioned that a spokesman for the satellite owner, TCI Satellite Entertainment, took a far more cautious position on why the satellite lost some of its operating power. Unlike Loral, which credited the space weather event for the problem, TCI announced that they were "unclear whether the storm had actually exceeded the levels the satellite was built to withstand, or if the satellite simply had a flaw." Compared to the murky causes surrounding the *Telstar 401* outage, this level of candor by Loral was refreshingly to the point, even though the satellite owner preferred a more guarded opinion. It would, of course, be the satellite owner that would seek insurance payments, not the satellite manufacturer. In a replay of the *Telstar 401* settlement, TCI filed a claim for $20 million.

Far from being just another satellite stamped from a tried-and-true design, *Tempo-2* was supposed to be the vanguard of a whole new fleet of high-capacity communications satellites. Communication satellites had evolved from humble 10-watt "small reptiles" to leviathan multi-kilowatt "dinosaurs" driven by the relentless evolutionary pressure of consumer demand. To generate the tens of kilowatts of power needed to operate dozens of transponders and other high-end equipment, engineers had been forced to create lighter-weight and higher-power solar cells. The current darling for this new technology is based on semiconductor compounds of gallium and germanium rather than the common silicon cell materials. The new cells would be wired to produce 60 volts per module to keep the weight and size of the solar panels within the limits set by the cost of the satellite. One of these panels, incidentally, could comfortably supply the needs of a medium-sized house. But the *Tempo-2* failure uncovered a potentially fatal problem with these new panels. They were susceptible to energetic particle impacts, which caused miniature lightning bolts to flare up and short-circuit sections of the panels. Engineers would certainly have to go back to the drawing board to fix this problem, because these satellites were to be the wave of the future.

Compared to winter and spring, the rest of the summer of 1997 unfolded in comparative calm. Although there is no public data on anomalies experienced by commercial or military satellites at this time, deteriorating space weather conditions by the end of September were openly cited as the cause of a Japanese satellite glitch. On September 20, 1997, the $474 million *Adeos* research satellite, launched by Japan a year earlier, began to malfunction. According to a report in *Space News*, "Cosmic rays were found to have damaged the main on-board computer, which caused it to shut down all non-essential systems, including the sensors, forcing scientists to reprogram its software." There was a low-level geomagnetic storm in progress near the Earth on this particular day, and in the weeks leading up to the malfunction *GOES-8* measured significant increases in "killer electrons" with very high energies. But there is no clear cause and effect to connect these events to the malfunction.

Close on the heals of the *Adeos* satellite problem, the Sun decided to get back into the act of terrorizing the Earth. The *SOHO* satellite witnessed two major CME events on September 24 and September 27. The events were echoed in the data returned by the *ACE* satellite on September 30 as the CME plasma rushed by the satellite at nearly one million miles per hour. *CNN News* and the *Reuters News Service* reported that India had lost an advanced communications satellite, *Insat-2D*, on October 2, 1997, because of a power failure. The satellite, launched June 4, 1997, carried twenty-four transponders for relaying Indian telephone and television traffic. The satellite's problems seem to have started on, or before, October 1, when it lost Earth lock briefly, shortly after the September 27 CME had passed the Earth on September 30. One of its predecessors, *Insat-1C*, launched on July 21, 1988, fell silent under similar circumstances in 1989. According to a summary for Insat-IC in *Janes Space Directory*, "a power system failure from a solar array isolation diode short" caused the satellite to lose half its capacity. On November 22, 1989, the satellite lost its Earth lock and was abandoned at a cost of $70 million.

For Insat-2D, *ACE* magnetometer data showed a sharp rise in solar wind strength on October 1 at 0000 UT followed by a persistent plateau of magnetic field intensity lasting a full day before subsiding again. The Earth-orbiting *Geotail* satellite also noted a sharp change in the local energetic particle conditions as well as geomagnetic field strength. All these are consistent with the arrival of the September 27 CME around October 1, at the time the *Insat-2D* began having its problems.

The *Insat-1C*, by the way, failed during the last of a series of major solar and geomagnetic storm events of this memorable year. These storms caused major increases in energetic protons, at energies above 10 MeV, between October 19 through November 6, rivaling all the events in Solar Cycle 22 taken together. On November 15 another powerful solar blast was detected on ground-level neutron monitors, and at the end of November another major solar flare, rich in high-energy protons, was recorded. According to space researcher Joe Allen at the NOAA National Geophysical Data Center, each of these events caused power panel degradation in a variety of satellites. Some lost five to seven years of usable lifetime as a result of the October proton events. Others suffered a variety of glitches and operational anomalies that were corrected by ever watchful satellite ground controlers.

After the loss of Insat-2D in October 1997, Thanksgiving, Christmas, and New Years passed uneventfully, at least by any outward appearance here on Earth. Meanwhile, 93 million miles away, the Sun continued to steadily ratchet up its tempo of delivering CMEs to nearly fifty per month. On April 15, 1998, there were some unsettled geomagnetic conditions here at the Earth, but nothing of great significance to space weather forecasters. Then, a few days later, on April 20 at 5:21 EDT, the strongest solar proton flare recorded since 1994 made a stellar appearance. It was an M1-class eruption, among literally a hand full of severe flares the Sun likes to cook up each cycle. In keeping with the unpredictable nature of these disturbances, even though the solar storm involved both a CME and a powerful solar flare, geomagnetic conditions here at the Earth were barely affected. A week later, a second round of storms commenced with a major outpouring of pyrotechnic events. This was, however, just a warm-up pitch for what would turn out to be one of the largest storms in 1998.

A series of low-level, X-class flares and multiple CMEs were expelled between April 29 and May 2. The outfall finally started to reach the Earth around May 2–3, causing a severe geomagnetic storm. Only three times before, during the January 1997, April 1997, and November 1997 encounters, had scientists gone so far as to provide formal press briefings for impending calamities. Like earthquake forecasting, it is better to miss a few quietly than announce false positives. But the conditions, this time, seemed to warrant some kind of official comment just in case the storms grew into something more provocative than the topic of a scientific research paper. So far, the scientists seemed to be batting a perfect game. The last three press releases had confirmed that scientists could anticipate the likely geospace impact of some solar storms. This new one would extend this winning streak to four.

On May 4 at 00:18 EDT a strong GIC affected the northeast United States as capacitor banks were tripped. This resulted in transformer saturation, which affected a major electrical substation in New Brunswick and caused voltage regulation problems throughout Maine. Minutes later, in New York state, voltages also started to drop in the eastern part of the state. The Nova Scotia electrical company measured 70 ampere GICs in one of its transformers. A day later, routine testing of a transformer in the Hudson Valley indicated insulation damage and a temperature spike of several hundred degrees Celsius. According to John Kappenman at Metatech Corporation, the Hudson Valley electrical utility recorded this problem to be due to "undetermined causes."

Satellite owners also experienced their own spate of problems, many of them fatal, between April and July of 1998. Apparently the first satellite to succumb to these conditions, and be publicly acknowledged, was the German research satellite *Equator-S*. On May 1 the satellite owned by the Max Planck Institute lost its backup processor. According to an announcement by the institute at the official web site for this satellite, "If a latch-up caused by penetrating particle radiation was the cause, there is hope that it may heal itself upon the next eclipse because of the complete temporary switch-off of the electrical system."

After no mention of any satellite problems among the seventy plus satellites now in orbit since their first launch on May 5, 1997, *Space*

News reported four Iridium satellites launched prior to May 1 had already failed. By May 8 this number jumped to five, and by July 23 seven Iridium satellites had failed or were ailing. One of the failures apparently had something to do with a satellite separation problem during a *Delta* 2 launch. The Iridium satellite network has been built with the assumption that about six satellites per year will have to be replaced. The year 1998 easily reached that mark. Although there is no hard evidence to suggest that the rest of the Iridium satellites failed because of the major space weather events of April–May, the timing of the press releases seems more than coincidental. So many in-orbit failures for a satellite system barely a year old led to a predictable loss of confidence on the part of stockholders. Investors voted with their feet by dropping the stock price for Iridium LLC ten dollars to 46.75 a share. Motorola's spokesman Robert Edwards noted, optimistically, in a *Space News* article that Motorola does not believe there is a common link behind the seven failures so far, although two of the satellites had their functions restored and are now working normally. According to *International Space Industry Report*, unnamed "industry sources" confided that "these are no doubt reaction wheel failures; at least seven wheels have mission-threatening problems or failures." Meanwhile, on May 7, another Big LEO satellite program, Teledesic, lost its vanguard experimental satellite. No details were given of either the cause of the malfunction, the systems involved, or the time the satellite failed.

The most spectacular outage since *Telstar 401* rocked the satellite community occurred on May 19, when the $165 million *Galaxy IV* satellite suddenly went for an unplanned stroll and mispointed its antenna. By some accounts, as many as forty million pagers in North America instantly went silent. This outage was followed on June 13 by the loss of the primary control processor on the *Galaxy VII* and an identical problem with the *DBS-1* satellite on July 4. PanAmSat Corporation, the owner of the three Hughes model *HS-601* satellites, was never able to identify a clear cause for these failures. *International Space Industry Report* carried a headline, "Hughes Hit Hard by Satellite Failures": "The failures have sent Hughes scrambling for an explanation, and left industry analysts wondering whether other Hughes-built satellites of the same family may be subject to similar problems."

Space physicist Dan Baker and his colleagues at the University of Colorado, however, uncovered evidence from NASA and NOAA satellites of a very disruptive space environment spawned by the April–May solar activity episode. There was a major NASA POLAR satellite anomaly on May 6, and more than a dozen anomalies plagued Japan's Global Meterological Satellite system between May 4–7. What was interesting about the data presented by Dan Baker was that it showed how active the geospace environment could remain even several weeks after a major CME impact. Could the outage of the Japanese *ETS-7* satellite two to three weeks after the November 7, 1997, storm fall into the same category of delayed satellite impacts? The March 22, 1991, flare was so powerful and rich in energetic particles that it actually caused a new radiation belt to form around the Earth. The Combined Release and Radiation Effects Satellite (*CRRES*), Geostationary Operational Environmental Satellite (*GOES-6*), and the Geostationary Meteorological Satellite (*GMS-4*) had no problems detecting this belt. The belt was even seen by dosimeters on board the *STS-40, STS-42* Space Shuttle missions and *MIR* space station up to ten months later. There were no space weather events at the time of the *Galaxy VII* outage, but *ACE* and *WIND* sensors detected a very strong disturbance in the solar wind between July 1–3. The more scientists study the response of the geospace environment, the more they seem to discover it is a morass of delayed effects and complex phenomena that don't always deliver their worst punches after an obvious, well-telegraphed encounter with the Sun.

As summer declined into the Labor Day hiatus, on August 27, 1998, a severe geomagnetic storm, and yet another press release, was triggered by the arrival of a CME event aimed directly at us. At 3:00 EST, on August 26, plasma from the CME finally arrived at the Earth and began to trigger a major geomagnetic response, recorded by the thirteen magnetic observatories operated by the U.S. Geological Survey as far south as Hawaii. The USGS, which seldom wades into the space weather arena in such a public manner, thought enough of this event to issue its own press release, "USGS Reports Geomagnetic Storm in Progress." Spectacular aurora, meanwhile, were observed as far south as North Dakota. Sometime during, or before, the week of September 9, another Iridium satellite suffered a malfunction.

September 23 saw yet another CME provide a replay of the severe "Labor Day storm." Again, dazzling green and red aurora, reaching as high as halfway up the northern horizon, shimmered in North Dakota and Canada. A major M2-class proton flare added to this chorus a week later on September 30.

The rest of 1998 passed uneventfully, with the Sun continuing to produce between 40–60 CMEs per month. The *SOHO* satellite, which had experienced a commanding problem on June 24, had by now recovered all of its scientific functions, and its handlers were learning how to operate it without the benefit of its gyros. Other satellite handlers were not experiencing such a happy conclusion to their labors, though. On December 20, 1998, the *NEAR* spacecraft was just beginning a crucial twenty-minute burn of its thruster to ease it gently into orbit around the asteroid Eros. The thrusters were turned on by the satellite following a prerecorded set of instructions, but suddenly the spacecraft aborted its firing. For twenty-seven hours, the satellite refused to speak to Earth, until ground controllers finally received a weak reply. They quickly uploaded commands for *NEAR* to take as many pictures as it could as it hurtled past Eros. Why had the carefully planned rocket firing gone awry in mid-execution? By June 1999, engineers had run numerous tests using identical computers and software but were unable to reproduce the glitch. A similar thruster firing had to be commanded exactly on January 3, 1999, so that *NEAR* could return for a second orbit insertion try in February 14, 2000. This time there was no glitch.

Not all mysterious problems in space necessarily have a space weather explanation as their root cause. Sometimes it really can be a finicky or missing line of computer code as NASA learned with the ill-fated Mars Polar Lander, or an outright hardware failure. These explanations usually become obvious after investigators carefully sift through the satellite's housekeeping data, which, like an airliner's "black box," describe what the satellite was doing before the mishap. Neither of these explanations seemed to apply to the *NEAR* spacecraft, since its second thruster firing went off without a hitch. But many problems can have something to do with local environmental factors, especially when we voyage into interplanetary space and confront the unknown. On September 29, 1989, a category X-9.8 solar flare was recorded on the backside of the Sun and not visible from the Earth. The *Magellan* spacecraft en route to Venus experienced power panel and star tracker upsets from the portion of the solar flare that had passed its way. *ACE* satellite observations of the solar wind show that it was far from calm around the time of the *NEAR* mishap. Even so, the specific environment in the vicinity of *NEAR* cannot be estimated from only *ACE* satellite data measured at one point in a different part of the solar system.

There were five significant solar proton flares recorded in January, April, and June 1999, but none strong enough to warrant much alarm near Earth orbit. The only significant satellite event to happen during the first half of 1999 occurred on March 12 when GE Americom's *GE-3* satellite suffered an anomaly affecting its station-keeping ability, causing the satellite to spin out of control. The satellite had been placed into service September 1997 and its transponders carried a number of feeds for CNN, PBS, and Turner Classic Movies. Readings from the *ACE* and *WIND* satellites at the time showed that, between March 10–13, there was a space weather event in the solar wind. Particle densities and magnetic field strengths increased five to twenty times above typical baseline levels before and after this event. There were no geomagnetic storms in progress on this date, however.

Once again, India's satellite weather service was brought low by the failure of another of their satellites, this time *Insat-2E*. A *Space News* report on November 29 mentioned that this satellite had started to have problems "about two months ago." Despite simulation studies, which showed that the undisclosed problem might be rectified, they finally went public with the problem and announced that the satellite was being taken out of service. Geomagnetic indicators provided by the CANOPUS and SESAME magnetometer networks in Canada and Antarctica showed that mild storm conditions existed around September 23 and 27, and *ACE* satellite data also showed two large space weather events at this same time.

The only really severe geomagnetic event in 1999 occurred on October 22, when conditions were elevated to a large storm status for a single day, the first time since November 1998. There were no published satellite outages, but John Kappenman notes that some North American power utilities recorded transformer trips on October 23 that could have been related to this storm.

Since October 1999, and continuing through June 2000, only a few serious storms have added their comings and goings to the story for this half of Solar Cycle 23. A major geomagnetic storm on April 7 pushed the Kp index to 8, only a few notches below the March 1989 storm. This time, a "classic" halo coronal mass ejection was spotted several days earlier by the SOHO satellite. The storm put on a substantial auroral display for Northern Hemisphere viewers and was visible from Tennessee, and even Florida. According to John Kappenman, there were a number of accompanying electrical problems from this storm as well. Between April

6–7, substantial GICs of 90–175 amperes were reported in transformers in New York and Pennsylvania. Several filter banks in Maine were tripped and had to be reset. There was a voltage sag reported in New York state, and in the Boston area the high voltage DC terminal on the Hydro Quebec tie into New England was temporarily affected. While North America experienced some problems, it was spared the main electrical impact of this storm at ground level. The highest GIC activity occurred between 19:00 and 20:00 EST on April 6, and the most intense portion of the storm was centered over Scandinavia. In Sweden one of the largest GICs ever recorded produced a 300-ampere surge in electrical equipment there. Curiously, there were no significant satellite anomalies reported. Once again, the physical phenomena playing themselves out above our heads seemed to be fickle and hard to anticipate. Following a two-month respite, between June 6–9 a series of M- and X-class flares and a CME caused significant radio disruptions with only a moderate Kp geomagnetic disturbance on June 8. This storm was covered by all of the major news networks and also turned out to be the inaugural event for the new NASA *IMAGE* satellite, which watched the ensuing complex flows of plasma and auroral activity from its orbital perch around the Earth. In a replay of the January 12, 1997, blackout, once again the Foxboro Stadium experienced a power outage on Tuesday night, June 6, this time during a U.S. Cup match between Ireland and the United States. For twenty-two minutes, a crowd of sixteen thousand fans sat in the darkness, bathed in torrential rains, while the problem was being fixed. No one was able to come forward with an explanation for the blackout, but the implication was that it was weather related. The *Washington Post* some days later announced a mysterious power outage on Monday, April 10, that managed to shut down Reagan National Airport at 7:50 P.M., causing numerous flight cancellations and prompting an FAA investigation. Localized power outages, of course, do happen somewhere in the United States and North America just about every day, so it's not clear that this one could be blamed on geomagnetic conditions. A more promising confluence of solar-terrestrial factors may have affected twenty-four thousand residents of a Chicago suburb, on Thursday evening, June 8, at 6:42 P.M. Once again, no obvious cause to this power outage could be found, although weather conditions on this 90-plus steamy day were quickly ruled out by Commonweath Edison. Meanwhile, this middling geomagnetic storm managed to do what its far more powerful predecessor in April had not been able to muster. In a very brief one-paragraph not in *Space News*, "Solar Flares Disrupt Two Commercial Satellites," some geosynchronous satellites did experience space weather-related problems, one on June 6 and the other on June 7, although the names of the satellites were withheld. Curiously, the date of the earlier satellite problem coincides with the Foxboro Stadium outage.

As you can see from the chronological listing of satellite outages and anomalies in table 9.1, the first half of this cycle has been far from inconsequential. In fact, it has afforded us as much excitement as many previous cycles have. The published satellite outages, totaling nearly $1 billion so far this cycle, do seem to happen close by some significant space weather event. The worst of these to date has been the April–May 1998 storm whose lasting effects seem to have left in its wake a number of satellite outages and severe operational anomalies. What is interesting as well from this record is that, with hundreds of operational communications and military satellites in orbit, the actual, severe satellite anomaly count from these storms is so small. We cannot know if this is simply the result of selective reporting, but, at least for the military satellites, it is obvious that military secrecy is a powerful inhibitor to announcing these satellite outages. The list in table 9.1 is probably complete, since it is difficult for major communication satellite lapses to be hidden from the news media when they occur. This suggests that, in fact, the vast majority of the present commercial satellites are surprisingly robust in operation, through even the most adverse space weather conditions we have seen since Cycle 23 began.

So far, Cycle 23 has brought with it a mixed bag of problems; the only thing missing is a major power blackout. No one really wants to benefit from the chaos that results from a severe blackout. Still, NASA space scientists and NOAA space weather forecasters would be silently grateful for a new "event" that would keep space weather forecasting in the public and congressional eye. Even engineers for the electrical power industry bemoan the lack of any significant storms during the first half of this cycle to trigger a Quebec-style blackout. Without a large socially significant event it is unlikely that electrical utility company managers will be interested in space weather forecasting and the latest GIC prediction systems. For the few thousand dollars this would cost per month, many electrical power managers are not convinced it would be worth the cost. Many of them worry about the more obvious sources of traditional power outages such as ice storms, tornadoes, and downed tree

TABLE 9.1 Published Satellite Outages and Severe Anomalies Between 1997–1999

Satellite outage or anomaly		Space weather event	
Date	Name	Date	Type
4/11/97	*Tempo 2*	4/11/97	M/G/A
9/20/97	*Adeos*	9/24/97	
		9/27–30/97	
10/1/97	*INSAT-2D*		
10/17/97	*PAS-6*		
10/28/97	*ETS-7*	11/7/92	M/A
		4/20/98	
4/28/93	*Iridium*		
4/28/98	*Iridium*		
5/1/98	*Equator-S*		
5/1/98	*Iridium*		
5.1.98	*Iridium*	5/2–3/98	M/A, Elcct. Kp = 5 +
5/6/98	POLAR		
5/7/98	*Teledesic 1*		
5/8/98	*Iridium*	5/6–19/98	
5/19/98	*Galaxy IV*		
6/13/98	*Galaxy VII*		
7/4/98	*DBS-1*		
7/23/98	*Iridium*		
7/23/98	*Iridium*	8/27/98	M/G/A, Kp = 7
9/9/98	*Iridium*	9/23–5/98	M/G/A, Kp = 6
		11/9–13/98	G, Kp = 5–6
12/20/98	NEAR	12/29/98	
3/12/99	GE-3	2/18/99	G, Kp = 6
		5/12/99	M/A
		9/13/99	Kp = 5
ca. 9/29/99	*INSAT-2E*	9/27/99	Kp = 5
		10/12/99	Kp = 5
		10/22/99	M/G/A Elect. Kp = 6
		4/6/2000	M/G/A Elect. Kp = 8
6/6/2000	unnamed	6/6/2000	M/G/A Elect. Kp = 5

NOTE: M = major CME event, A = major aurora, G = Geomagnetic storm. Elect. = electrical power grid event, Kp = geomagnetic disturbance index and maximum value recorded.

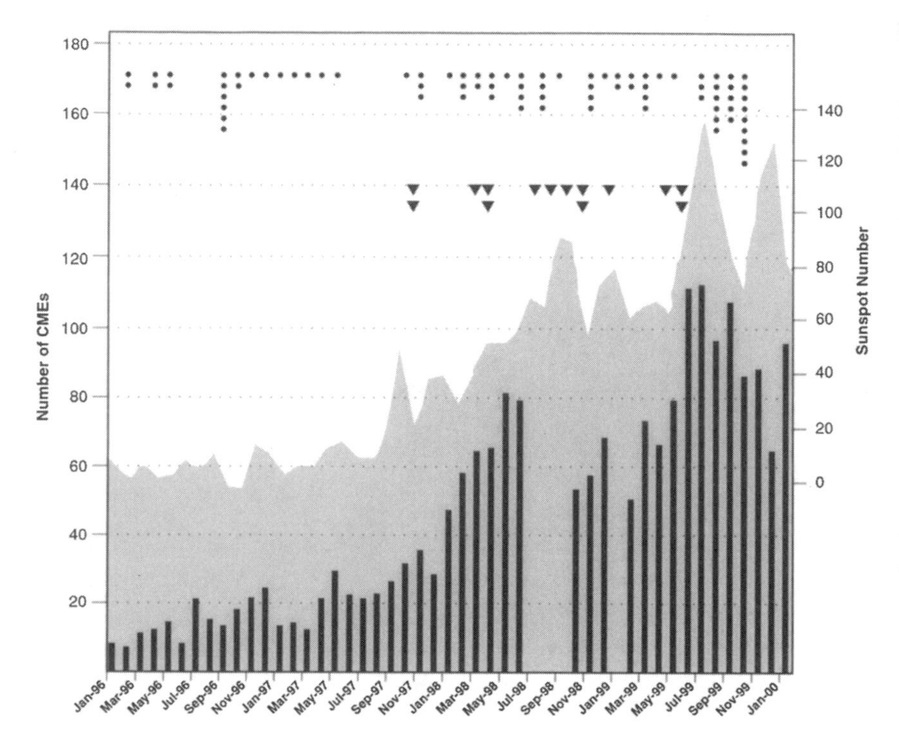

FIGURE 9.2 A summary of the first half of Cycle 23. Bars indicate the number of CMEs each month detected by *SOHO/LASCO*. Shaded region indicates monthly averaged Zurich sunspot number. Filled circles indicate geomagnetic storms with $Kp > 3.0$. Triangles represent significant solar proton flares detected by *GOES* spacecraft. The gap in the CME record between June 1998 and October 1998 is due to the temporary loss of the *SOHO* satellite.
SOURCE: *SOHO/LASCO, NOAA/NGDC*

limbs. But the absence of a compelling event hasn't been because the Sun has, somehow, kept a low profile.

Between June and September 1999 alone, one hundred CMEs per month blasted out from the Sun into the dark interplanetary depths, fully three times the rate that the current solar cycle had started with back in 1996. The minor blips that the lucky few squalls delivered to our shores did little more than cause a steady drum beat of geomagnetic storminess and a few good aurora: great for amateur photographers, bad for low-visibility space weather forecasting. We still have another three to four years to go before solar minimum conditions are reinstated. If the first half of this cycle has been an expensive one for satellite insurers, the second half may still harbor some surprises.

As Cycle 23 pushes beyond its peak and begins its descent during the next few years, we still have to consider it a cycle to watch carefully. It will, after all, be during the declining years of the cycle that much of the planned space real estate will be in place and serving our needs. Some of the most medically and technologically troubling storms have a penchant for happening during the years immediately following sunspot maximum. Astronauts will be assembling the International Space Station and inhabiting it in what many hope will be "routine" shifts. Everyone is counting heavily on the fact that space station crews won't have to be rushed back to Earth to avoid being radiation poisoned. Shuttle missions continue to play a cat-and-mouse game with the LEO environment, but the handful of days of accumulated spacewalks between 1996–2000 have not provided much of a target for the Sun's storms. Since Cycle 23 began, and by the end of June 2000, satellite and ground-based instruments have steadily ticked off about eighty significant geomagnetic storms, fifteen solar proton flares, and an unbelievable two thousand plus CMEs. Ultimately, we have to remind ourselves that the game we are playing isn't chess, where pieces always move logically and in recognizable patterns. Instead, the game is largely one of chance. There are, however, many ways that we can improve the odds that our technology will survive unscathed. One of these is simply to become much more adept in anticipating when solar storm conditions will occur and what their specific impacts will be here at the Earth when they arrive.

Part III

The Future

10 Through a Crystal Ball

> Two significant flares occurred during August 20–21 as a historically active sunspot group returned to the visible face of the sun. The geomagnetic field was disturbed through August 20. The source of the disturbance was a high-speed solar wind stream that originated from a coronal hole on the sun's surface. Spacecraft sensors detected solar wind speeds approaching two million miles per hour. There's a chance for more significant solar flares from the sunspot group during August 25–31 as it continues to trek across the visible face of the sun.
>
> —NOAA/SEC, "Outlook 99-20," August 24, 1999

When CMEs do make it to the Earth, the compressed magnetic fields and plasma in their leading edges smash into the geomagnetic field like a battering ram. Across a million-mile-wide wall of plasma, the CME pummels the geomagnetic field. Such niceties as whether the polarities are opposed or not make little difference to the outcome. The CME pressure can push the geomagnetic field so that it lays bare the orbits of geosynchronous communication satellites on the dayside of the Earth, exposing them to wave after wave of energetic particles. When the fields are opposed, particles from the CME wall invade the geospace environment, amplify ring currents, and generally cause considerable electromagnetic bedlam, often tracked by increases in the recorded satellite anomalies and power grid GICs. Clearly, we need more advanced warning for solar flares, geomagnetic storms, and CMEs. A successful forecast of how severe a particular solar cycle will be, no matter how accurate, is simply not much more than a statement that "this winter will be more severe than last year." That isn't enough information to prepare us for tomorrow's snowstorm, so is there any way of doing better than just predicting the ups and downs of the next solar cycle? Is there any way we can get the jump on individual day-to-day solar storms and space weather events? With some effort, the answer is, luckily, "Yes," but, like a Trojan horse, there are actually three kinds of forecasting issues tucked away within this single operation. You can attempt to predict a space weather event before it starts. You can try to predict what it will look like when it is en route to Earth. Or you can predict what it will do when it arrives.

If we can watch the Sun, we can gauge when a CME will come our way and often have two or three days advanced warning. For solar flares, on the other hand, there is still a lot of work to do to provide more than a ten-to-thirty-minute warning before they erupt on the solar surface, and we still can't predict just how powerful the flare or the stealthy proton releases will be when they get to the Earth. This means, for astronauts, that every flare sighting requires running for cover from the X rays as if your life depended on it. Of course, by the time your instruments register a problem, it could well be too late.

Once solar physicists had studied solar flares for a long enough time, they began to develop a scale for ranking their magnitude. Originally, it was a crude optical scale, but then came satellites equipped with X-ray and proton detectors. The oldest scale measures the flare's X-ray intensity. This scale is actually rather fluid during the sunspot cycle. During sunspot minimum, the X-ray brightness of the Sun is low, so a flare of a given brightness can be quite spectacular, like a flashlight in a dark cave. But during sunspot maximum, when the Sun is far brighter as an X-ray source, this same flare is nearly invisible, like a flashlight switched on in broad daylight. There are four main classes in increasing order of strength: B, C, M, and X. Each category is broken into ten numerical subcategories: 0–9. An M5.5-class flare, for example, is ten times more powerful than a C5.5-class flare and one-tenth as powerful as an X5.5-class flare.

The second scale, recently adopted by NOAA for its space weather alerts, ranks flares on the basis of their energetic particle flows measured at the Earth. An S2 flare has ten times the particle flow of an S1 flare, but the classification of the flare is based on the actual number of par-

TABLE 10.1 Quick Guide to Space Weather Indices

Class	Description	Impact	Strength	Frequency
S5	Extreme	Astronaut EVA lethality. Satellite loss, star tracker errors	100,000	1 per cycle
S4	Severe	Astronaut EVA hazard, satellite anomalies	10,000	3 per cycle
S3	Strong	EVA not recommended. Satellite anomalies. Component damage.	1,000	10 per cycle
S2	Moderate	No biological problems. Few satellite upsets.	100	25 per cycle
S1	Minor	No problems	10	50 per cycle

NOTE: The strength of the flare is in units of particles per second per square centimeter per steradian.

ticles, not an electromagnetic intensity as for the B-, C-, M-, and X-classes.

Big Bear Solar Observatory is a telescope located on a spit of land in the middle of a lake in Southern California. Unlike most observatories perched high up on the mountaintops, this one was placed on a lake because of the peculiar stability and clarity of the solar images that result from the combination of geographic circumstances. Air turbulence is normally the biggest factor preventing astronomers from seeing small details on the Sun. At Big Bear Lake, the air flows are parallel to the water and help to reduce the amount of turbulence near the telescope. Harold Zirin and William Marquette have spent years perfecting their BearAlert Program for spotting solar flares before they hatch. Armed with real-time solar data, they watch the minute-to-minute changes in an active region traced by its magnetic field and hydrogen emission. As a public test of their methods, over the course of a two-year period, they issued thirty-two "BearAlerts" for sizable flares via e-mail and scored hits on fifteen of them. Because solar conditions generally do not include flares, and because the Sun's state changes only very slowly from day to day, it is possible to issue a "no flare today" warning and be correct nine times out of ten. This promising score is, of course, useless for anticipating whether a flare will actually happen or not. BearAlerts, and the space weather reports they have evolved into, are issued only when a flare seems about to happen. They are the closest things we have today to keeping ahead of these unpredictable solar storms.

Weather forecasters can usually tell you whether a particular storm has what it takes to unleash lightning discharges over your city during a given two- to six-hour period, but that is their limit. In a similar vein, solar physicists are fast approaching the ability to announce that a given active region will spawn solar flare activity during a set six-day period but, ironically, can't tell you if one will happen in the next few hours. Like weather forecasters, they can't tell whether you will get a few major flares that could affect astronaut health or a hail of minor flares that, individually, are unimportant.

The next element of space weather is the solar wind itself, which acts like something like a conveyor belt, connecting the surface of the Sun and activity there with the Earth. After the spectroheliograph was invented in the 1890s, astronomers quickly got an eyeful of fiery prominences, and other phenomena, busily hurling matter into the space surrounding the Sun. But no one appreciated just how far this star stuff could travel until clues to its invisible journey began to show up in the direction that comet tails pointed and in direct spacecraft observations in the early 1960s. It travels at speeds of about a million miles an hour and has a density of about ten to fifty particles per cubic inch, mostly electrons and protons. In fact, it's a better vacuum by far than what scientists can make in their laboratories. What makes this wind disproportionately complex compared to the breezes you feel on a summer's day is that it carries a magnetic field along with it.

When two magnetic systems such as the Sun and the Earth interact, the outcome depends on whether the polarities are the same or are opposite to one another. If the wind and geomagnetic polarities are the same on the daytime side, the geometry dictates that they can cause the solar wind to slide over the outskirts of the Earth's magnetic field and flow smoothly into the depths of space. The interaction of the north-type geomagnetic field with a south-type solar wind field, on the other hand, is usually very dramatic. Hours-long geomagnetic storms and spectacular aurora result, as currents of accelerated particles flow from distant unstable regions in the dynamic magnetotail and into the atmosphere along the field lines. North- and south-type magnetic field lines rage a pitched battle to unkink themselves into a smooth geometric shape. As a result, the magnetosphere picks up energy from the currents of particles that are created and the geomagnetic field becomes wildly unstable in its outer frontiers: the magnetopause. Because the origin of these magnetic storms involves the invisible solar wind whose roots in the solar surface cannot be detected, they seem utterly random and unrelated to specific sunspot groups. We never see them coming. Milder storm conditions can be spawned as the wind constantly changes its strength and polarity. The geomagnetic field responds to these changes with magnetic irregularities called "substorms." Substorms last a few hours, but are sometimes strong enough to cause aurora to appear in extreme northern and southern latitudes. Even comet tails develop kinks and irregularities that follow the clumpy, and gusting, solar wind.

About one million miles from the Earth, in the general direction of the Sun, a group of NASA satellites serve as our outposts on the solar wind at the L1 Lagrange point. The L1 region is an invisible dimple in the gravitational well of the rotating Earth-Sun system. You could fly right through it and not realize anything unusual was going on. Satellites carefully positioned there, like a pencil balanced on its point, may orbit this invisible point in space, lacking any gravitating matter to hold them to this spot. From this vantage point, *SOHO* busily watches the solar surface and relays its images back to Earth. The *ACE* satellite, meanwhile, samples the magnetic field and composition of the solar wind as it rushes by. Like buoys bobbing in the ocean off coast, these satellites tell us of changes in the wind conditions that can signal trouble for the geomagnetic field within forty-five minutes.

In addition to solar flares and the solar wind, the coronal mass ejections, first seen by the *OSO-7* satellite and by *Skylab* in 1973, have been studied in detail, and nearly all of them vouch for a serious consequence should one find its way to the Earth. Soon after being launched by the Sun, in an event that from our vantage point on Earth often engulfs nearly the entire solar disk, they are accelerated to speeds from a gentle 10 km/sec to over 1500 km/sec–nearly two million miles per hour. Within a few days, they can make the journey from the Sun to Earth orbit and can carry up to fifty billion tons of plasma.

The launch of the *SOHO* satellite in 1995 put the Sun under a twenty-four-hour weather watch. One of the most spectacular instruments on this satellite was LASCO, the Large Area Solar Coronal Observatory. Like its predecessors on *OSO-7* and *Skylab*, it was a coronograph, which manufactured artificial total solar eclipses so that the faint details in the corona could be studied. No sooner had the shutter opened on this instrument, when it began to record vivid images of CMEs leaving the Sun. Within a year, *SOHO* scientists became adept at using LASCO to anticipate when the Earth would be affected by these disturbances. Eventually, NOAA's Space Environment Center, whose responsibility was to produce daily space weather forecasts, began to use the LASCO data in 1996 to improve their accuracy. By keeping an eye out for "halo" CME events that were directly aimed at the Earth, it was now a routine matter to achieve a two- to three-day advanced warning at least for the onset of major geomagnetic storms that could cause satellite outages and electrical power blackouts. So long as the *SOHO* satellite keeps working, it substantially improves our chances of never being caught off guard the way we were during the Quebec 1989 blackout.

Although solar flares are often seen near the birthplaces of CMEs, solar physicists don't believe they are what actually cause them. CMEs and flares both track yet more subtle underlying conditions that are probably the mother to them both. Flares actually happen at much lower altitudes in the Sun than where the CME plasmas are spawned. Solar physicist Richard Canfield and his colleagues at the University of Montana have spent some time trying to get the jump on CMEs even before *SOHO*'s instruments can start to pick them up. They think they have found what is triggering at least the major ones that we have to worry about back home. To see the birth of a CME, you can't use ground-based data at all. You have to use X-ray images of the Sun taken by satellites such as the Japanese-U.S.-British Yohkoh X-ray Observatory.

A major press briefing at NASA Headquarters on March 9, 1999, soon got the news media's attention, and the *Washington Post* carried a headline, "Scientists Find Way to Predict Solar Storms," while *ABC News* offered, "The Sun's Loaded Gun: **S**-Shapes on Surface Foretell Massive Solar Bursts." The idea that these **S**-shaped "sigmoid" fields were like a cocked gun ready to fire became the inevitable centerpiece sound bite in many of the reports. During sunspot minimum, about one CME can be produced each day or so. During sunspot maximum, the Sun can spawn a handful of them in a single day. Fortunately, most of these are ejected either on the opposite side of the Sun from the Earth, or at large angles from the Earth so that they miss us about nine times out of ten. When CMEs flare toward the Earth, Great Aurora bloom across the globe, and geomagnetic conditions become dramatically turbulent for days as the great wall of plasma rushes by.

> Strong geomagnetic storm conditions are in progress. These levels of activity are possibly the result of a shock observed in the solar wind on October 21 at 01:38 UT originating from a coronal mass ejection on the sun on October 18. This level of disturbance routinely causes power grid fluctuations, increased atmospheric drag, and surface charging on satellites, intermittent navigation system problems, signal fade of high-frequency radio signals, and auroral displays at mid-latitudes. (NOAA/SEC Advisory 99–9, October 22, 1999)

The geomagnetic field and its collections of trapped particles is the last stop for most of the Earth-directed severe space weather events spawned by the Sun. Just as your local weather reporter can tell you about rainfall, temperature, humidity, and pressure as presages for tomorrow's forecast, space weather can also be charted by keeping track of a handful of numbers. Over the years, scientists devised a number of quantities that gave a quick reading to the level of geomagnetic storminess. Few have turned out to be as popular as the Kp index devised in 1932 by Julius Bartels. In addition to counting sunspots as a barometer

TABLE 10.2 Geomagnetic Storm Indexes Used in Space Weather Forecasting

Scale	Level	Impact	Kp	Frequency
G5	Extreme	Blackouts. Aurora seen at equator	9	4 per cycle
G4	Severe	Transformer trips, pipeline currents. Aurora seen in tropics	8	100 per cycle
G3	Strong	Electrical false alarms. Aurora seen in mid-latitudes	7	200 per cycle
G2	Moderate	High-latitude power grids affected. Aurora seen in Canada	6	600 per cycle
G1	Minor	Weak power fluctuations. Aurora in N. Canada	5	1700 per cycle

of solar activity, the Kp index brings a second dimension to the problem of forecasting: sunspot numbers define how active the Sun is, while Kp tells how vigorous the Earth's geomagnetic response was to solar activity, or to other phenomena, that can disturb the Earth's magnetic field.

Kp is a measure of the largest swings in magnetic activity that you record around the globe during any three-hour period. It's not a number on a linear scale like temperature; instead, it's a part of what is called a semilogarithmic scale. A Kp = 9 geomagnetic storm, for example, is about five times stronger than a Kp = 6 storm. Typically, on any given day, the Earth's field imperceptibly bumps and grinds at Kp levels between 1.0 and 3.0. With a magnetic compass in hand, you would not even know there was a problem at all. These seemingly random gyrations define the normal quiet state of the planetary field, but occasionally it can belt out a disturbance you need to pay attention to. Kp values between 4.5 and 5.5 are classified as small storms like the occasional, harmless, earthquakes seismologists detect every few weeks in the San Francisco Bay area. Large storms require Kp values between 5.6 and 7.5, and these are analogous to the yearly shakes California residents feel that cause the dishes to rattle and the chandelier to swing. Finally, you get to the major "head for the hills" storms that require Kp indices greater than 7.5 and resemble the once-in-a-decade Loma Prieta or 1999 Turkey earthquakes. They are the ones that can cause blackouts. Luckily, geomagnetic storms have to be pretty large before anyone has to seriously worry about what immediate impacts they will have. Only storms with Kp indices greater than about 6.0 seem to have what it takes to shake up electrical systems. On this scale, the Quebec blackout was a 9.3 "mega storm." There have only been three other ones like it in the last fifty years: in 1940, 1958, and 1989. With that said, space scientists cannot tell you when the next one will happen. One thing is for certain, based on previous patterns: the odds are very high that there may be less than a few minutes warning that the storm will escalate to this level of severity—not enough time for a utility company to do much more than watch and hope for the best. By the time you are forced to use Kp to decide what to do, it is already too late to decide what to do.

So, after one hundred years of research, space physicists have now begun to understand some of the basic rules of space weather forecasting. They know how to measure a set of parameters that track space weather severity. They have at their disposal real-time images of the solar surface and its surroundings. There are many parallels with ordinary weather forecasting, too. Like modern weather forecasters watching a hurricane develop, they can track CMEs as the leave the Sun, but they lose sight of them almost immediately as they enter interplanetary space. Fortunately, just as hurricane watchers on a beach can see an incoming storm hours before it arrives, satellite sentinels at the L1 Lagrange point can anticipate a CME shorefall on Earth within the hour. Meanwhile, solar physicists can anticipate when an active region on the Sun may disgorge a flare, but, like weather forecasters, they cannot predict the times of individual lightning strikes.

Unlike terrestrial weather forecasting, however, the main problem that opposes the further development of newer space weather forecasting techniques is that the data are too sparse to follow all the changes that can have adverse impacts. Research satellites are launched and put into service on the basis of scientific needs, not on the basis of their utility to space weather forecasting. Only NOAA's monitoring satellites and their military equivalents are specifically designed to serve space weather forecasting needs. But, even if we had a fully working armada of satellites keeping watch on the entire system, this would still not be sufficient to provide detailed forecasts. Some method has to be found for filling in the data gaps, and that method involves the detailed physical modeling and measuring of the system and all its various interactions.

In ordinary weather forecasting, scientists have thousands of stations throughout the globe that report local temperature, pressure, humidity, wind speed, and rainfall. Weather balloons and rockets as well as satellite sensors measure changes in wind speed and pressure across great swaths of vertical space from the ground and into the tropopause. Every minute or hour, a "state of the atmosphere" survey can be made to poll how things are going. To make a forecast about tomorrow's weather, you plug this data into a sophisticated 3-dimensional model, which extrapolates the current conditions into the future, one small computation step at a time. It's called the general circulation model, and it is the product of a century's work in the scientific study of the atmosphere using the tools of classical mechanics, thermodynamics, and the behavior of gases and fluids. When you mix these theoretical ingredients together with the data on a rotating, spherical surface heated by the Sun, and connected to the oceans and land masses, the resulting atmospheric model helps the National Weather Service generate forecasts good enough to make the average person happy. The one-hour forecast is usually bang-on correct. The twenty-four-hour forecast is now routinely accurate for perhaps ninety-five attempts out of one hundred. The three-day forecast is usually good to about seventy attempts out of one hundred, unless you live in Boston–where nothing works. Even the seven-day forecast is better than the toss of a dice in many localities. Weather forecasts are also more accurate the larger the area they apply to. For instance, you may not be able to predict the rainfall in Adams, Massachusetts next Wednesday, but you can tell if El Niño will make the entire East Coast of North America warmer or cooler by two degrees. With long-term climate models, you can even recover the global weather patterns for the spring of A.D. 769.

Now, suppose you only had a dozen weather stations across the globe, and every five or ten years you had to replace some of them at a cost of $150 million each. Suppose, too, that when you replace them you don't put the new ones in the same locations or equip them with the same instruments. You also don't get to make the measurements at the same time. Then, added to this, suppose that your forecasting model is still under development because you don't know what all of the components that affect your weather happen to be. You don't know how clouds move from place to place, or how the sunlight actually heats the gas, or just what it is that causes rain to form in a cloud. Welcome to the complexities of space weather forecasting:

> Solar activity, between December 1–27, is expected to range from low to high levels. Frequent C-class flares are likely. Isolated M-class flares will be possible throughout the period. There are also chances for isolated major flares as potentially active regions 8765, 8766, and 8771 are due to return on December 7. There is a chance for a solar proton event at geosynchronous orbit when the above mentioned regions return starting on December 7. The greater than 2 MeV electron flux at geosynchronous altitude is expected to be at moderate to high levels during December 5–10 with normal to moderate levels during the remainder of the period. The geomagnetic field is expected to be at unsettled to minor storm levels during December 4–8 due to recurrent coronal hole effects. Otherwise, activity is expected to vary between quiet and unsettled levels barring any earth-directed coronal mass ejections (NOAA/SEC Weekly Highlights and Forecasts, December 1, 1999, 2112 UT)

By the 1980s, solar and geospace research had made a number of significant refinements to the best of the theoretical models for how the space weather system functions; much of this was thanks to the advent of powerful supercomputers and new data from dozens of interplanetary observatories and spacecraft. Everyone could now afford their own "workstation" that harnessed more computing power than most of the mainframe computers of the 1960s era. What was dramatic about the new way for researchers to do business was that it was no longer necessary to take mathematical shortcuts that could compromise the accuracy of a theoretical prediction. Nearly photographic renderings of complex fields, plasma flows, and particle currents could be calculated and compared to satellite data as it was taken along the satellite's actual orbit. Theoretical investigations were now hot on the trail of being able to describe the detailed bumps and wiggles in satellite data, not just their overall shape. Because the calculations were based on "first principles" in physical science, they were powerful numerical testing grounds of our knowledge of the space environment. Glaring deficits in understanding tended to show up like a black eye, impelling theorists to improve the mathematical models still further. The art of modeling space weather systems had matured to the point that the crude averages used in earlier AE-8 and AP-8 models which NASA had developed during the 1970s were no longer necessary or even desirable.

The next big challenge was to combine a number of separate mathematical models into one seamless, coherent, and self-consistent supermodel. The National Weather Service had long enjoyed the benefits of a general circulation model to predict the course of a hurricane or next Tuesday's rainfall. What space weather forecasters needed was something very much like it. During the 1980s, researchers independently worked on their own theoretical approaches to space weather phenomena, each describing a specific detail of the larger system. In the 1990s, it was time to bring some of these pieces together. Here's how it is meant work, at least in principle:

In the new scheme of things, a solar surface "module" developed by one group of researchers would take a set of input conditions describing the solar surface and calculate the surface magnetic conditions of the Sun along with the various plasma interactions and flows. This information would be passed on to a solar wind, or CME, module developed by other groups, which would detail the transfer of matter and energy from the solar surface all the way out to the Earth's orbit. At this point, you would have a forecast of whether the Sun was going to send a CME toward Earth or not.

The output from this solar wind module would then feed a geospace physics module, which would calculate the detailed response of the Earth's magnetosphere, ring currents, and magnetotail conditions. Finally, there would be an upper atmosphere module that would take the output from the geospace physics module and calculate how the properties, currents, energy, and composition of the Earth's exosphere-ionosphere-mesosphere system would be modified.

Like a relay race in which a baton is passed from one research team to another, a disturbance on the Sun would be passed up the stack of modules until a specific consequence materialized in the geospace environment. Each of these steps would be updated in near–real time for a "Nowcast" or jumped forward five, ten, forty-eight hours to make extended forecasts based on the current conditions. At least this was the hope. In reality, although the individual parts to the "car" were in hand, there was no agency that could assemble all the parts. No single agency had the financial resources and scientific support to do it alone. The Department of Defense (DoD) might, for instance, have the best available model of the ionosphere. The National Science Foundation (NSF) and NASA might have supported research to develop the best available solar atmosphere model. The knowledge had to be shared and interconnected before it would be possible to make a meaningful forecast. This requires the cooperation of scientists working in many disciplines under many different kinds of grants, across a number of different federal and private agencies.

Even though space environment effects have been known for decades, space weather forecasting is nearly as much an art as a science. By some accounts, we are forty years behind the National Weather Service in being able to detect or anticipate when a solar storm will actually impact the geospace environment and what it is likely to do when it arrives. Meanwhile, the Weather Service has benefited from two critical developments during this same time frame. Powerful "physics-based" programs have been created that run on supercomputers to track atmospheric disturbances from cradle to grave. This is possible because our theoretical understanding of what drives atmospheric disturbances has grown and deepened since 1950. The second factor is a functioning network of weather satellites, which actually watch the globe around the clock and have done so almost continuously since the early 1960s, when *Tiros* was first placed in orbit. All this atmospheric research and monitoring activity is supported by NOAA's National Environmental Satellite,

Data, and Information Service, which maintains a fleet of polar-orbiting and geosynchronous weather satellites to the tune of $368 million (FY 1997) a year. Some of these satellites, such as the *GOES* series, even carry space environment monitors. There is no comparable network of nonresearch satellites to keep track of space weather conditions.

Only in the last five years have scientists been able to put in place a ragtag collection of satellites capable of keeping constant, and simultaneous, watch on the solar surface, the solar wind, and its effects on the geospace environment. Although NASA has launched more than sixty research satellites since the early 1960s, studies of the space environment are still regarded as low-profile activities compared to planetary exploration and probing the deep universe. The need for a specific satellite is weighed entirely on its scientific and technological returns to NASA and the space science community, not on any benefit to NOAA or commercial and military space weather applications. This is an attitude very much different than for weather satellites such as the *Tiros*, *GOES*, and *NOAA* series launched by NASA but operated by NOAA. There are dozens of these applications satellites orbiting the Earth that are owned by non-NASA agencies, like NOAA, the Department of the Interior, and the Department of Defense, compared to a handful of working research satellites.

As the twentieth century began to draw to a close nearly forty years after the start of the Space Age, members of the space science community thought that it was a good time to start thinking about the big picture. So in 1993 they went ahead and contacted the National Science Foundation. In response, NSF called for a meeting of government, industry, and academic representatives to discuss what was going on in space weather research and what kinds of things needed to be done. The federal coordinator for meteorology was assigned the task of organizing this huge program which would take quite some effort to set in motion. It was pretty obvious, by then, that several decades of independent work by researchers in many agencies still had left, nevertheless, many things only partially completed in terms of a larger product such as a space weather forecasting model. Like tiling a floor, sometimes it is easier to work at the center of the floor than in the complex boundaries. But some invisible threshold had been crossed, and everyone agreed that the new National Space Weather Program (NSWP) would be worth the cost. According to NSWP, "The predominant driver of the program is the value of space weather forecasting services to the Nation. The accuracy, reliability, and timeliness of space weather specification and forecasting must become comparable to that of conventional weather forecasting."

NSWP would have to work with such diverse federal agencies as the NOAA, NSF, DoD, and NASA, all having long, historical ties to different segments of the research community and with their own needs for improved forecasting capability. The DoD, for example, has its own space weather service provided by the Air Force's Fiftieth Weather Squadron in Colorado Springs, and they share in operating the SEC at NOAA in Boulder. Their particular interest is how solar and geomagnetic storms affect the LORAN navigation system, Global Positioning System satellites, and other sensitive satellite real estate. They had one of the best ionosphere models in the world, but were understandably concerned about secrecy issues in just handing over the model's computer code and operating theory to the non-DoD community.

To start the ball rolling, NSF and the DoD made $1.3 million available in 1996 to augment space weather research in several key areas and promised to increase this amount each year. NSF added this new research directive to its Global Change Research Program through a new initiative called Geospace Environment Modeling. The outcome of this research would be a geospace general circulation model that would take solar wind conditions and forecast their consequences for the entire geospace region. A series of "campaigns," begun in 1996, would support theoretical modeling grants for researchers to study the magnetotail region and how it causes substorms and the inner magnetosphere with its ring currents. This sounds like a lot of money, but, in reality, nearly half of the $1.3 million per year will disappear into various forms of institutional "overhead" costs including phone bills, office space rental, and health benefits. Out of hundreds of space scientists, only a few dozen or so will be supported each year on this kind of a budget to do the herculean job of building this mammoth space weather modeling system. But it was a far cry from no support at all! By FY 1999, this amount had increased to $2 million, and the NSF was hoping to use this to support twenty to thirty scientists at $50,000 to $100,000 per year, including overhead costs.

NASA already supported much of this activity through its Office of Space Science, which handles Sun-Earth Connection research. NASA's role in space science has by no means been inconsequential. Since 1958 it has built and launched over sixty solar and space physics research satellites at the behest of the space science community. With Congressional approval, NASA created satellite programs such as *Explorer* and *MIDEX*, that paid teams of researchers to build the instruments and the satellites. NASA then launched these payloads. Afterward, NASA provided all the satellite tracking and data archiving services for the duration of the funded mission. Each mission has a budget for Mission Operations and Data Analysis (MO&DA) from which it supports its own investigators to work with satellite data. NASA also hires its own permanent staff of space scientists to support the archiving activities and provide modest enhancements to the format of the data so that the space science community can work with the data more efficiently. Ironically, NASA space scientists and mission scientists cannot apply to the National Science Foundation to support their research. NSF does not support space research using NASA resources. NSF considers any research involving space or satellite data something that NASA should support. Moreover, NASA rarely supports astronomers to carry out ground-based research involving telescopes. NASA "Civil Service" scientists, meanwhile, can only conduct research that enhances the value of the satellite data. Although mission space scientists are sometimes offered permanent jobs with NASA when no hiring freezes are in effect, they usually return to academia or industry and continue their research, sometimes by obtaining both NSF grants and NASA research grants.

Beginning in 1996, NASA's Office of Space Science tried to set up a Quantitative Magnetospheric Predictions Program that was supposed to result in a comprehensive magnetospheric model. The model would rely on solar wind data provided by its own research satellites such as *WIND* or *ACE* and from this compute the consequences for the complete system. It was a promising and exciting new program, and a timely one to boot, but the idea never became a funded NASA program. The message from Congress, and from NASA, to the scientific community was that NASA had already done its fare share of contributing to the National Space Weather Program just by *providing* the research community with satellites and data. Any work that NASA's space scientists would do with the archived data would have to focus on providing "value-added" information, not producing a major product such as a new forecasting model. At the request of the non-NASA research community, NASA had put into place a virtual armada of solar and space physics research satellites, and NASA was very happy to supply non-NASA modelers with all the data they needed. After forty years, there was a lot of data to go around.

At the NASA Goddard Space Flight Center, Building 28 is tucked away in a not very well-traveled part of the campus. Deer frequently come out on the front lawn to graze and keep a wary eye out for passing scientists. The 1990s vintage architecture hides a virtual rabbit's warren of offices and cubicles, each with its own occupant hunched over a computer terminal or reading the latest journal. It is also the home of the National Space Science Data Center, a massive, electronic archive of all of the data obtained by NASA satellites since the early years of space exploration. Satellites numbering 395 have contributed 4,400 data sets and a staggering 15 terabytes of data that grows by 100 gigabytes each month. There are also 500,000 film images from the manned space program, and hundreds of movies and videos.

Sophisticated, interactive programs such as the Consolidated Data Analysis Web (CDAWeb) let scientists extract specific measurements of dozens of different physical properties that define space weather conditions throughout the solar system. You can do this too, if you visit their Internet page! Would you like to see what the solar wind magnetic field was like on January 1, 2000? Enter the date, select the magnetic parameter, and in a few seconds you will get a plot of magnetic field directions from the *ACE* or *WIND* satellites. A little more of this data mining will quickly point out a problem. There are big gaps, in both space and time, in the available data for a given parameter you are looking at, because satellites and their measuring instruments have not been flying at the same time to perform coordinated studies of specific phenomena. This lack of coordinated observations began to change in the early 1990s with the International Solar-Terrestrial Program: ISTP.

This $2.5 billion program, inaugurated in 1994, used the vast majority of this money to build four key satellites and to support engineers and other ground crew to keep round-the-clock vigils on spacecraft functions and telemetry. The Solar and Heliospheric Observatory (*SOHO*) monitors the solar surface at optical and ultraviolet wavelengths to catch CMEs and keep watch on active regions on the Sun. *WIND* measures the solar wind speed and magnetic field strength at the L1 Lagrange point inside the orbit of the Earth. Next in line is the *Geotail* satellite whose complex orbit lets it measure activity in the magnetotail of the Earth, watching for changes that herald the onset of geomagnetic substorms. Last, the *POLAR* satellite looks at the polar regions of the Earth to keep watch on the changes in auroral activity.

In principle, this fleet of satellites can study the cradle-to-grave growth of solar disturbances and track them through a series of satellite handoffs all the way from the solar surface to the auroral belt. The ISTP network has only been in place since 1996, which means that it hasn't been "on the air" long enough to examine a representative number of solar storm events. In fact, it started its campaign during sunspot minimum when not much was going on at all. Although in 1998 there were some plans to stop funding ISTP at NASA, by 2000 this prospect seems to have vanished, and NASA is now committed to fully funding the ISTP program at least until the satellites themselves begin to fail. *SOHO* and *POLAR* have already had their share of technical problems, and *SOHO* was nearly lost for good during the summer of 1998. Although the funding now seems to be stable, there are real concerns that the satellites themselves will not survive much beyond the peak of Cycle 23, a critical period for catching the Sun at its worst.

Since ISTP became operational, NASA has also provided an array of other satellites beyond the ISTP constellation as new technology and scientific interests arose. By 1998, the Sun, the wind from the Sun, and the geospace environment have been under around-the-clock surveillance by a newer generation of satellites. None of these missions, however, have a carte blanch to do more than a modest amount of research with their data before archiving it for posterity.

The Advanced Composition Experiment (*ACE*) satellite, launched in 1998, monitors the minute-to-minute changes in the solar wind magnetic field and composition. This $160 million mission hopes to retain NASA funding until its steering gases run out in 2006. Despite the many, and growing, practical benefits of having this satellite operational until the end of Cycle 23, it faces stiff competition from other planned research satellite programs to continue operating beyond 2001. NASA, and the space community, is less interested in practical benefits from a satellite than a steady stream of fundamental insights about space physics processes. The predecessor to *ACE*, called *ISEE-3* and launched in 1978, ran into similar difficulties. NOAA and DoD wanted this satellite to remain at L1 to continue providing real-time solar wind data for space weather forecasting. NASA, at the urging of its science advisory board, yanked it out of this location so that it could flyby Comet Jacobi-Zimmer in 1983. The Air Force made it quite clear to NASA that *ISEE-3* was needed for practical purposes, but NASA had to listen to the science community that sponsored the mission to "explore" and do a pre-Halley's

Comet flyby. *ACE* currently costs $5 million each year to maintain the satellite and to fund research scientists to work with, and archive, the data. Again, NOAA and the Department of Defense, not wishing or being able to secure the funds themselves, rely on NASA to develop and launch satellites, like *ACE*, to help with their space weather forecasting.

The Transition Region and Coronal Explorer (*TRACE*) satellite, launched in 1998, uses high-resolution imaging to show the fine magnetic details on the solar surface that older satellites such as *Yohkoh* could not detect clearly. The promise of better advanced warning for CMEs, and especially for solar flares, will be realized by the crystal-clear images returned by this satellite of magnetic field structures on the solar surface. Even grade school students will study these dramatic images to learn about solar magnetism. The $150 million mission will last until 2003, with no currently planned replacement to continue the exploration of the solar magnetic "fine structure."

The exciting prospect of actually imaging CMEs as they travel from the Sun will become a reality in 2001 with the launch of the Solar Mass Ejection Imager (*SMEI*). This satellite, developed by the U.S. Air Force's Battlespace Environment Division at the Air Force Research Laboratory, will measure sunlight scattered by electrons within the CME and create movies of incoming CMEs. Extensive studies by Bernard Jackson, the University of California, San Diego coinvestigator on the *SMEI* mission, has already demonstrated how well this technique works using data from the *HELIOS* satellite in 1977 and radio-wavelength data from ground-based telescopes. As a forerunner to the next generation of CME imagers, it will almost completely take the guesswork out of predicting which CMEs, out of the several thousand the Sun produces every sunspot cycle, will actually collide with the Earth.

Closer to home, the geospace environment will not be left out of this onrush of investigation. The $83 million Imager for Magnetosphere-to-Auroral Global Exploration (*IMAGE*), launched in March 2000, provides images of nearly the entire geospace region to keep track of the movements of charged particles and their currents. Previous generations of satellites only measured the space weather conditions where they were specifically located. This is like trying to track a hurricane by only using scattered weather stations in Florida and South Carolina. *IMAGE* replaces this kind of data taking by imaging nearly the entire contents of the magnetosphere cavity. This will revolutionize the study of the magnetosphere in the same way that the first weather satellites photographed and tracked hurricanes from space. Like conventional weather satellites, *IMAGE* delivers five-minute update images of the global pattern of plasmas, from the magnetopause all the way down to the auroral region. For the first time, space physicists will be able to "see" the flows and changes in these systems of particles that previous satellites could only hint at. The satellite's prime mission lasts two years, with a much hoped for extension until 2004, assuming that the space science community continues to see this satellite as actively contributing to magnetospheric research. What *IMAGE* scientists hope to learn from this is how high energy particles circulate and are stored in the magnetosphere, which will then tell space scientists about the latency of energetic particles. In practical terms, it may also illuminate how satellites such as *Galaxy IV*, *DBS-1*, and others, sometimes seem to run into trouble long after a space weather event has seemingly passed us by.

In the first decade of the twenty-first century, a new series of NASA satellites such as *STEREO*, the Global Electrodynamics Connections, and the Magnetosphere Multi-Scale Mission will replace the current fleet. An ever changing hat game will be continued as older satellites run out of fuel or funding and have to be replaced by newer, more capable, satellites designed to explore new issues in the Sun-Earth system.

After the ISTP program disbands as its satellites, one by one, fall out of service from old age, what new program will take its place to coordinate another assault on the space weather issue? The current suite of satellites is mostly a series of independent efforts led by investigators studying specific issues, but there is only a rudimentary attempt at coordinating the observations. In some cases, it is not possible to do this because, for example, a satellite like *IMAGE* may not live long enough to be on the scene when the *STEREO* satellites begin taking their data. *IMAGE* will rely on a, hopefully, one to two year overlap with *SOHO* and *ACE* to provide data on the external, interplanetary environment that sets in motion the geomagnetic events *IMAGE* hopes to investigate. But the key problem is that there is not enough research money outside the satellite operating budget to support scientists in making sense of what they observe. To make matters worse, over the years, the part of a mission's budget that is set aside for research, MO&DA, often gets robbed during the construction of the satellite to cover cost overruns. One solution is for NASA to create a program, with more available money to go around, to support both new satellites and enhanced MO&DA activities. In 1996, NASA attempted to create the Quantitative Magnetospheric Predictions Program and ISTP. Although the former program did not survive as a new start, ISTP succeeded spectacularly and provided a coordinated investigation of solar activity during the first half of Cycle 23. In 1999, NASA proposed another program to take over from ISTP and to further coordinate space research activities.

> The ultimate output of this campaign would be the observational specifications for an operational space weather system and the models to apply to the data to produce accurate and reliable forecasts over the timescales required to be beneficial to humanity's space endeavors. (NASA, SEC 2000 Roadmap, p. 96)

Every three years, federal agencies are required to develop strategic plans to serve as a basis for governmental policies and strategic planning. In January 2000, George Withbroe, the director of NASA's Office of Space Science, together with a team of twenty-eight experts, produced the "Sun-Earth Connections 2000 Roadmap." A significant factor in this document is the renewed emphasis placed on improving our space weather forecasting ability and providing the satellite resources to keep a constant watch on the Sun through the year 2025. Withbroe's new program, which he calls *Living with a Star*, is the embodiment of the new strategic plan and will nearly double the $250 million spent on solar and geospace research each year by NASA. With the backing of his advisers from the space science community, he envisions a new suite of satellites to be built in the first decade of the new millennium that will take over from the aging ISTP program and cover the next solar cycle: Cycle 24.

In August 1999, following an unusually lengthy meeting with NASA's administrator, Daniel Goldin at NASA Headquarters, Goldin gave his go-ahead to Withbroe's proposal to set up such a new program, and since then Withbroe has been presenting his plan to the scientific community to galvanize support for it. Apparently, it wasn't the detailed science or the heroic dreams of solar physicists that apparently caught Goldin's attention. Instead it was an issue, in the post-*Challenger* NASA age, that has become a critical ingredient to every scientific program administered by NASA: safety. Astronauts can, and will, be affected in a measurable way by radiation exposure. Even though the Occupation and Safety Administration and NASA have agreed upon the 50 rem per year annual limit for astronauts, in today's radiation-averse society, even this much (equal to thousands of chest X rays) seems an unacceptable health risk. Some solar flares can do far worse than this dosage to a spacesuited astronaut. In a press release by the National Research Council issued on December 10, 1999, they also urged NASA to carefully monitor its astronauts for radiation exposure and to support programs that will enhance our ability to forecast solar storms. Newspapers such as *USA Today* carried the story, originally covered by the Associated Press, with the headline "Radiation Alert": "[The NRC] warned that astronauts might receive doses of radiation equal to several hundred chest X-rays from solar flares during planned space construction."

Although *Living with a Star* is an exciting new program with profound impacts on space weather forecasting, it still has to meet the challenges of another, even larger program, "Living with the Congress." NASA may recommend a "new start" program requiring a new "budget line" to be opened in NASA's annual budget, but it literally requires an act of Congress to make it happen. Although we enter the new millennium with over $200 billion in federal budget excesses each year, NASA's own budgets are projected to be extremely flat for the foreseeable future, making it very difficult to shake loose the money needed for a new program. Coming as it does as a new proposed expense for NASA during an election-year congressional budget debate in the year 2000, the odds seem pretty slim that *Living with a Star* will reach ignition temperature. Nevertheless, a rumor has it that sometime in late 1998, while NASA was testifying before Congress, the issue of what NASA was doing about space weather came up in the questioning of NASA's planned FY 1999 budget. If true, this could be a watershed moment for the future of this entire enterprise at NASA and a promising sign that its time has, at last, arrived.

> Solar storms are dramatic changes in our solar system that are the result of solar activity. The ground doesn't shake, and the sky does not turn black when a solar storm strikes the Earth. . . . Because solar storms attack the very foundation of our high-tech society, scientists are excited to find that satellite data will help them predict solar storms and mitigate their impact on Earth. ("Our Sun: A Look Under the Hood," NASA Facts)

More than just another NASA program that will benefit NASA and the academic space science community, one of the major beneficiaries of this new program will be the Space Environment Center in Boulder, Colorado. This will happen in the same way that the U.S. Weather Service benefited from the atmospheric research spurred on by the new satellite data provided by NASA in the 1960s. The mission of the SEC is to conduct research on solar-terrestrial physics, develop techniques for forecasting geophysical disturbances, and provide real-time monitoring of solar and geophysical events. The fifty-five employees that work there under a $5 million annual budget issue daily forecasts to a long, and in many cases confidential, list of clients including the U.S. military and commercial satellite owners. Whether you are a global positioning system (GPS) user, a geologist prospecting for minerals, or even a pigeon racer, you may find yourself in need of one of these forecasts to avoid bad conditions that could cost you time and money or get you lost. The modest annual budget for the SEC expended to create these forecasts is insufficient to build its own space weather satellites. It also seems an astonishingly small investment given that over $110 billion in satellite real estate and hundreds of billions of dollars of Gross Domestic Product can be impacted by space weather events.

This promise of substantially improving our space weather forecasting capability, both scientifically and practically, may be stillborn as Congress wrangles over an important technical issue. In May 2000, the House of Representatives deleted NASA's fledgling *Living with a Star* program from the Fiscal Year 2001 budget. They were concerned that this program sounded too much like an active space weather forecasting system of its own and not a pure research investigation. NASA's mission is to do research, not to delve into practical applications. In defense, NASA replied that Congress is giving this agency mixed messages by also instructing the agency to foster more applications-oriented commercial research. *Living with a Star* was supposed to be the vanguard of this new wave of thinking. It is hoped that funding for the program will be reinstated by the Senate and survive the August–September congressional budget negotiations to follow. During the 2000 election year, there are many pressures being placed upon Congress, and space weather issues may seem far too theoretical and abstract to carry the necessary political support. Still, we have to remain hopeful, because the cost of doing nothing has begun to grown unacceptable with each passing solar cycle.

Because of a lack of data, and a regular stream of it that scientists can count upon over time, our understanding of space weather is still primitive. We cannot anticipate so much as a day in advance what solar region will spawn a solar flare or a coronal mass ejection. We cannot anticipate the properties of a CME with any reliability until it reaches one of NASA's ageing sentry satellites (*ACE, WIND, SOHO*) in L1 orbit, one million miles from the Earth. This gives us barely thirty minutes to recognize a problem is on the way. Satellites such as *POLAR, IMP-8*, and *Geotail* patrol geospace but cannot be everywhere at once to give us literally a ten-second warning. With resolutions measured in thousands of miles, we cannot anticipate how the geospace environment will respond to a storm at a level of detail that is useful for a specific military or commercial satellite. Instead, many spacecraft designers have to rely on statistical models of the geospace environment that are thirty years old. This is like trying to predict tomorrow's rainfall in New York City using data from the same day of the year averaged between 1960 and 1970.

It isn't just the satellite industry and NASA's manned space program that will benefit from the next generation of forecasting tools provided by *Living with a Star* and the National Space Weather Program. The third leg of this particular stool is the electrical power industry. Progress in this area has been difficult because of the widespread opinion that an electrical power emergency caused by adverse space weather is so infrequent that it is ignorable. In fact, this is not the case at all, as we discovered in Chapter 4. Every time there is a geomagnetic storm with a severity of Kp = 6 (a G3 event), electrical utility companies in the northern-tier states experience strong GIC currents that trip some of their protection systems and require manual intervention to reset them. When Kp reaches 7 or 8 during a G4 event, dozens of these temporary interruptions sweep across the electrical grid of Japan, North America, Scandinavia, and Great Britain. When Kp reaches 9, (a severe G5 event) as it does at least once every solar cycle, hundreds of equipment failures sweep across North America and Europe in a matter of a few minutes. Depending on the time of year and the amount of operating margin available, blackouts become an expensive and public reality.

PEPCO, VEPCO, and BGandE, despite their locations in regions that are usually not greatly at risk from geomagnetic storms, are no strangers to outages. A January 1999 ice storm turned out the light and heat for over four hundred thousand people in Maryland, Virginia, and Washington, D.C. for up to five days. Although it was not widely reported, it was a major hardship for many residents of the Washington, D.C. area. The electrical utilities were under constant unrelenting attack from private citizens and the media to reconnect their services. One street waited hopelessly by for two days, while the lights on the streets surrounding were quickly brought back on. The bad press and harsh feelings directed at the electrical companies undid years of hard work by the utilities to portray themselves as "friends." It was not surprising that these same utilities appeared at a conference in Washington, D.C where Bill Feero rolled out his Sunburst system and asked them for support. Even rare geomagnetic events could throw their customers into a frenzy, and the few thousand dollars for the Sunburst system seemed like a bargain.

Meanwhile, an electrical utility company running John Kappenman's PowerCast system can look at any line, transformer, or other component in their system and immediately read out just what it will do when the solar wind hits the Earth traveling at a million miles per hour. With thirty minutes to spare, it is now possible to put into action a variety of countermeasures to protect the grid from failure. PowerCast has just become operational in Great Britain, where it has been used for several years to improve the reliability of their national power grid. Entry into the North American utility system has been sluggish because, at a cost of a few thousand dollars per month, many utility managers still do not see it as a high priority investment given that space weather disruptions are so infrequent.

Ironically, the ACE satellite seems constantly on the verge of cancellation by NASA to make way for newer missions. The fact that ACE data plays such a vital role in GIC forecasting for the power industry seems to be of no special interest to NASA. NASA is, after all, a research organization supported by the US taxpayer, not a for-profit corporation looking for commercialization opportunities. The viability of the ACE mission at NASA hinges totally on its scientific returns and not its potential for practical applications. NASA also has to make way for future missions with the declining, and politically vulnerable, space research budgets that the U.S. Congress, in its wisdom, has mandated. In England, which uses the PowerCast technology, ACE is seen as an ally in keeping their entire multibillion dollar electrical utility system operating reliably. Rutherford Appleton Labs has invested in its own independent ACE satellite tracking station to intercept the solar wind data. Arslan Erinmez, chief engineer at the National Grid Company in England, notes that "The British power industry would be happy to do anything it can to keep ACE going." While the destiny of satellites such as ACE turns completely on how well its scientists can convince NASA and Congress not to terminate it, its politically silent, commercial clients both domestic and foreign continue to mine its data to help the power industry keep your electricity flowing.

NASA, and the space scientists that advise this agency, are not interested in building a follow-on satellite to *ACE* just to supply private industry with a forecasting tool, unless it can be justified on solely scientific terms of advancing our understanding. Even so, any prospective follow-on to *ACE*, such as the *Triana* satellite, will have to compete with astronomy missions such as the Next Generation Space Telescope to secure its funding, and with MAP, AXAF and Hubble Space Telescope to maintain their year-to-year operating budget. NASA has been forced into a zero-sum, or even declining, fiscal game by Congress, at a time when space research has exploded into new areas and possibilities. Whether the power industry gets a GIC forecasting tool to keep Boston lights turned on, or NOAA's Space Environment Center can help satellite owners prevent another major communication satellite outage, hinges on whether at NASA investigating quasars and the nature of dark matter is deemed more important than studying the physics of solar magnetic field reconnection.

Epilogue

> Typically, engineers are becoming more specialized and are less likely to understand the hazards of the space environment. Often we have to relearn things that were known by people who retire or move on. . . . It is also recognized that private companies have a disinclination to release information on their own problems.
>
> —NOAA/SEC Satellite Working Group, 1999

Sometimes we work too hard to make coincidences into real cause and effect: recall the example of the *Exxon Valdez* accident during the March 1989 space weather event. And sometimes it's not easy to grasp just how complex space weather issues have become in the last ten years. There are many facets to the story and, like a diamond, the impression you get depends on your perspective. When I first started learning about this subject, I was overwhelmed by the lack of careful documentation, and the impossibility of ever finding it, for many of the outages I had heard about. I was also nervous about the basic issue of simultaneous events masquerading as cause and effect. If you are intent on seeing every mishap as a demonstration that satellites are inherently vulnerable, then there is ample circumstantial evidence to support the claim.

Why is it that satellites are assumed to be unaffected by events like cosmic rays, solar flares, or energetic electrons until such a claim can be "proven" by recovering the "body"? If we know that space is a hostile environment to off-the-shelf and nonradiation hardened systems, why do multinational corporations greet the inevitable failures with suspicion when they do happen? Where did this presumption of innocence begin to enter the professional discussion?

So long as there is any scientific uncertainty about cause and effect, well founded or otherwise, there will be no measurable change in the attitudes of satellite manufacturers. As NASA Goddard space scientist Michael Lauriente notes, "Because of the statistical nature of the risks, apathy reigns among some of the spacecraft community." Space physicist A. Vampola from the Aerospace Corporation also confronts the irony of the sparring between scientists and satellite owners, "The space environment is hostile. It is not a benign vacuum into which an object can be placed and be expected to operate exactly as it did when it was designed and tested on the ground."

There are plenty of uniquely interested parties who would denounce each satellite "kill" as simply a technological problem with insulation or some other factor, and there is also ample cause to support this point of view. Scientific data is too sparse for us to look into every cubic yard of space and say, with authority, that "this satellite was killed by that event." This means that you are completely free to argue that satellites are actually quite invulnerable to any space weather effect, and that the failures are purely a matter of quality control. But, if that is the case, why, then, do satellite owners take out insurance policies? Why do they overdesign solar panels and go on record saying that they expect to lose up to six satellites per year?

A satellite communications corporation may very well be intent on maintaining the status quo in supporting the ragtag effort of space weather research–and may find just cause to believe that this strategy is OK, too. After all, severe solar storms capable of producing a blackout happen very infrequently compared to other kinds of power outages (ice storms, line sags, and the like). Since satellite owners suppress evidence of anomalies traceable to solar storms, no trail of problems remains to require better space weather models. However, NOAA has many satellite owners on confidential lists that receive daily updates about space conditions. In a classic catch-22 situation, these clients would like better forecasts but cannot support such an effort at NOAA or NASA without going public and implying that their service is vulnerable.

While satellite owners and electric power utilities continue to expand their vulnerability, and reap billions of dollars in annual profits, space weather forecasting continues to languish. Both NASA and NOAA have been forced into "flat" or declining budgets for these specific activities. Back in June 1997, NOAA's Space Environment Center was facing a funding crisis with flat funding or zero growth planned for the years following 1997. According to Ernest Hildner, the director of NOAA's Space Environment Center in Boulder, Colorado, "Without increased funding, we will be facing the next solar cycle with two-thirds as much staff as we had during the last solar cycle." A similar route of the National Geophysical Data Center was attempted in the 1970s.

Space physics, and modeling research in support of space weather forecasting, tends to be underfunded given the complexity of the subject. Through much of the 1990s, efforts to develop better forecasting models at NASA were also stuck on the slow track. This occurred at the same time new floods of data entered the NASA data archives. Space physics doesn't seem to have a constituency the way cosmology and planetary exploration does. A mission like SOHO costs $1.2 billion to build and support, of which $400 million was supplied by NASA and the rest provided by the European Space Agency. According to Art Poland, who was the former NASA project manager and project scientist for SOHO,

> SOHO can support 20–25 scientists full time, and has done so for the last 4 years. Including institutional overhead costs, it costs about $150,000 to support each scientist. The total research budget for a mission like SOHO is about $8 million per year. To support the entire ISTP program each year including technicians, engineers and scientists costs about $50 million.

So, despite the billions of dollars that go into building and supporting satellite hardware, in some cases less than a few percent of the total mission cost ever shows up as support for scientists to analyze the incoming windfall of new data. This means that, in some instances, less than $50 million per year supports scientists to develop the forecasting techniques to safeguard $100 billion in satellite assets and $210 billion in electrical power company profits. Researchers may get dazzling images of CMEs from one satellite like *SOHO* between 1996–2001, solar wind data from another satellite like *ACE* or *WIND* from 1999–2002, and geospace data from another satellite such as *IMAGE* between 2000–2002. This information may not overlap in time–it may not even overlap in wavelength–yet it all has to be knitted together. This makes developing space weather models a frustrating enterprise, because it is easier to decipher the language of space weather if you have complete sentences rather than fragments.

No agency other than NASA is mandated to build the satellites that keep watch on space weather. This means that space weather monitoring activities are forced to piggyback on satellites optimized for pure research. The organizations that benefit most from NASA research satellites receive the space weather data free of charge and, for a variety of internal reasons, are unable to make a significant financial investment in data acquisition themselves. This solution is certainly easier than investing in space weather satellites, but it also means that newer, better forecasting techniques, for example, in the electric power industry, are held hostage to the necessities of ongoing yet uncertain NASA funding.

The National Academy of Sciences evaluated the readiness of NASA, NOAA, DoD, and DOE to develop a comprehensive space weather forecasting model. Although they identified many successes in the existing programs, they also identified glaring problems that work against accomplishing the National Space Weather Program goals anytime soon. NASA, up until 1997, was planning to stop funding its ISTP program after fiscal year 1998; NOAA's Space Weather Center suffered a 33 percent staff reduction and had stopped supporting the translation of data-based models into forecasting tools. "The lack of commitment at NOAA to this unique and critical role will have a fundamental impact on the success of the NSWP." A similar critique was leveled against DoD, where the staff turnover time was less than a single eleven-year solar cycle so there was no institutional memory of what was learned during past cycles.

Fortunately, NASA's *Living with a Star* program, which was conceived in 1999, promises to keep ISTP fully funded until its spacecraft expire from wear and tear. The program will also set in motion a more aggressive support of space weather research during the next decade. To be ready for the next solar maximum in 2011, missions must be planned today so that the necessary hardware will be in place when the new round of solar activity rises to a climax.

Beyond the interests of the research community, many different elements of our society have recently begun to appreciate the need for a deeper understanding of the space environment. The military and commercial satellite designers, for example, are not happy that they have to use twenty- to thirty-year-old models to predict space weather conditions. This forces them to fly replacement satellites more frequently or overdesign others to withstand even the occasional major solar flare, which may never come at all. Unfortunately, the current trends in satellite design seem to be directed toward increasing satellite vulnerability to disabling radiation damage. As pointed out by William Scott in *Aviation Week and Space Technology,*

> Austere defense budgets also have increased reliance on more affordable, but perhaps less robust, commercial off-the-shelf hardware. . . . Expensive radiation-hardened processors are less likely to be put on some military satellites or communication systems now, than was once the case according to USAF officers. . . . Newer chips are much more vulnerable than devices of 10–15 years ago.

In an age when "cheaper, faster, and smaller" drives both civilian and military satellite design, satellites have become more susceptible to solar storm damage than their less sophisticated but more reliable predecessors during past solar cycles. As more satellites become disabled by "mysterious" events that owners prefer not to discuss openly, old lessons in satellite design need to be rediscovered. In recent advertisements, some satellite manufacturers boast that they employ "advanced composite materials which improve performance while reducing weight." This also makes for poor shielding because weight, and quite a bit of it for shielding, has proven itself to be a good radiation mitigation strategy.

In all this concern over satellite survivability, there looms another harsh circumstance that may make long satellite life times impossible. Large networks of satellites in LEO are suffering something of a shakeout that has nothing to do with the rise and fall of solar storms. The pace of communications technology, and consumer needs, has begun to change so quickly that by the time a billion-dollar satellite network is in place it is nearly obsolete. Companies have to install their networks within one to two years or run the risk of becoming a technological backwater before the planned life span has been reached. The two companies that have put in place the two largest networks by 1999, Iridium and Globalstar, have already filed for bankruptcy. Iridium was able to sell only ten thousand of the one hundred thousand wireless telephone handsets it planned, and these phones did not have the capability to send data along with voice transmissions. No sooner had Iridium gone on the air than its owners regretted the decision not to include high-speed data channels. Globalstar, meanwhile, had not even completed installing its full satellite system before it followed Iridium LLC into bankruptcy. Yet this kind of record, so soon in the fledgling LEO industry, doesn't appear to bother some people. Satellite communications is still seen as a highly lucrative business despite lightweight satellites that will inexorably face space weather problems. The tide of technological progress is sweeping the industry at an ever faster pace. A sense of urgency now pervades the industry: make profits as quickly as possible. For example, Bruce Elbert, senior vice president of Hughes Space and Communications International, and the leading manufacturer of geosynchronous communications satellites, suggested with great enthusiasm, "The next millennium may see all land-line communications going wireless. You could wait [to enter this market], but why put off gaining the economic and potentially competitive advantages of using satellite technology today?"

Meanwhile, communication technology has not stopped evolving. New devices and systems are on the rise that may eventually make communication satellites less desirable, at least by the peak of the next solar cycle in 2011.

The first fiber optic cable, TAT-8, entered commercial service in 1988, and since then no fewer than 408,000 miles of fiber have been laid across the oceans by 2000. The current investment in undersea fiber exceeds $30 billion and is expected to surpass $50 billion by 2003. One project alone, Project Oxygen, will be a $14 billion cable tying thirty countries in Europe and North America together. It will take no more than three months to lay this cable and, when it is completed, it will carry twenty-five million phone calls and ten thousand video channels. It can carry all the world's Internet traffic on one of its four fibers, which can deliver 320 gigabytes per second of capacity. The combined bandwidth of 1.2 terabytes per second is enough to transmit the entire text contents of the Library of Congress in a few seconds. Even so, the signals carried by these fibers have to be amplified, and these electronic units can be vulnerable to space weather effects on the ground and under the ocean.

The driving force behind the spectacular growth in fiber optic technology is the Internet and the insatiable appetite it creates for massive volumes of data delivered immediately. At the same time, the explosive growth in the market for cellular phones has driven the satellite industry to meet traveler's desires to stay in touch with family and coworkers no matter where they may be on the Earth. The biggest drawback to fiber optic communications is that to take advantage of the high data rates it requires land lines to individual users. Satellites, meanwhile, require that

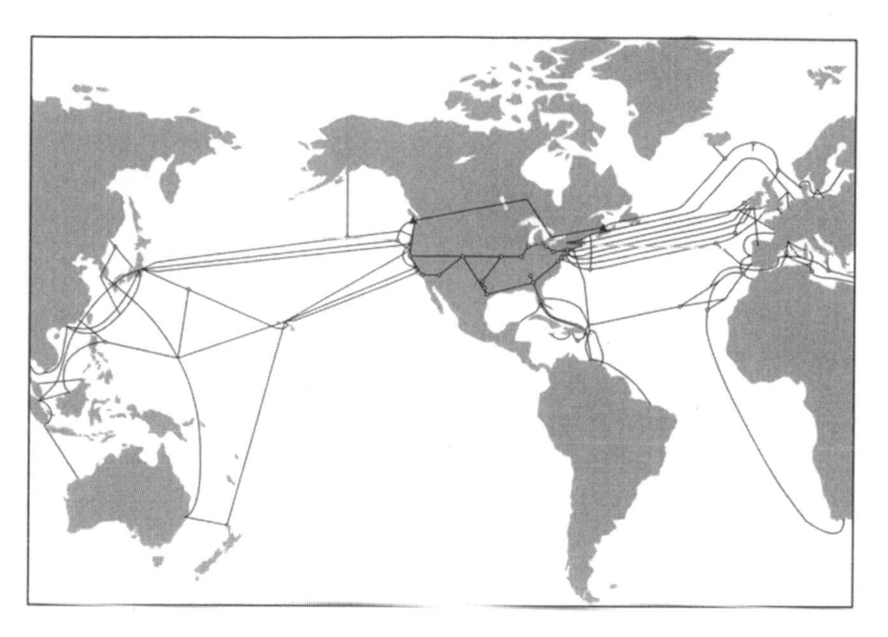

FIGURE 11.1 Current and planned undersea fiber optic cables to be in place by 2003.

their users need only a portable handphone or a satellite dish to receive their signals directly. Satellites work well when connecting large numbers of rural or off-the-road users. Yet fiber optic cables still have the advantage of the highest data rates, and they do not have to be replaced every five to ten years the way satellites must. Modern cables are not rendered useless as new, faster, technologies arise. Only the equipment at the cable's end stations has to be upgraded, not the entire undersea fiber cable. In addition, when a cable is damaged, it can be easily repaired underwater using remote-controlled robots. On the other hand, satellite transponders create the bottleneck problem as the capability of this technology increases, and they can only be upgraded by launching a new $200 million satellite.

The pace of fiber optic cable evolution and deployment has been so brisk that even the U.S. National Security Agency has been forced to rethink its intelligence operations. With more and more sensitive data being carried by fiber, the NSA's powerful satellite eavesdropping systems no longer hear as much of the secret messaging that they used to pick up from satellite or radio transmitters.

Unlike satellites, which largely support our entertainment and financial needs, the pervasive use of electrical power can create far more serious problems than even the most dramatic satellite outage. Electrical power outages can completely shutdown an entire region's commerce for hours or days, and they can cause death. Geomagnetic storms will continue to bombard the power grid, and there is little we can do to harden the system. The blackouts we will experience in the future will have as much to do with our failure to keep electrical supply ahead of demand as it does with the failure of the electrical power industry to put in place the proper forecasting techniques. Unlike satellite design, there are few easy solutions to the declining operating margins because power plants seem to be universally unpopular. Our vulnerability to the next blackout has become as much a sociological issue as a technological one.

When the ice storm of January 1999 darkened the lights of nearly half a million residents in Virginia, Maryland, and Washington, D.C., no one thought it was especially amusing or trivial, even though only 0.1 percent of the U.S. population was involved. My family and I were among the last to have our electrical service restored after waiting in the cold and dark for five days. The first night was a curious mix of concern and genuine delight at the new experience. We huddled together under down comforters and actually sang campfire songs as our home slowly gave up its latent heat. We were delighted to see the beautiful trees, like sculptures of ice, bent over along the street, which was now an impassable skating rink. Elsewhere, emergency rooms were filling up with people who had broken bones, turned wrists and ankles, or blood streaming down from head wounds and concussions. Hundreds of traffic accidents offered up a handful of fatalities as people died for no reason other than being in the wrong place at the wrong time. By the third night, we had joined thousands of others in Montgomery County, calling the electrical utility company to find out when we might be reconnected. We found ourselves the only street in our community that did not have electrical service. At night, we watched our neighbor's security light cast ghostly figures of tree limbs on our bedroom walls, by day we frequented local shopping malls to keep warm and ate our meals at restaurants. We felt a temporary kinship with homeless people who endure this daily. It wasn't the Sun and its mischief that had brought this on, but a common natural occurrence of more mundane origin. Still, the discomfort, expense, physical pain, embarrassment, and even death that it caused is a potent reminder that we can no longer afford to be as tolerant of power and communication outages as we once were. When you gamble with space weather, a major winter storm can occasionally happen at the same time as a large solar storm.

When I first bemoaned how astronomy and astrophysics seldom have practical consequences, I hardly suspected that simply the search for answers to questions in cosmology and galactic astronomy could cause a domino fall that could make blackouts more likely. The very satellites that I backed as a professional astronomer to further my particular area of research and curiosity were in the zero-sum game of modern budget analysis, robbing the space scientists of the tools they would need to improve space weather forecasting. In the future, the next power outage my family and I have to endure may, in some sense, be the consequence of my own professional choices elsewhere in life.

Our vulnerability to solar storms during Cycle 23 follows a pattern that can be seen in other more familiar arenas. Between 1952 and 1986, the average cost of national disasters (earthquakes, tornadoes, hurricanes, and floods) averaged $8 billion every five years. This then doubled in 1987–1991 to $20 billion, and in 1992–1996 the cost skyrocketed to over $90 billion. The reason for this sudden increase in the last five to ten years is that more people, often with higher incomes, are moving in droves to coastal communities–exactly those areas typically battered by earthquakes and hurricanes. As Greg van der Vink and his colleagues at Princeton University's Department of Geophysics have noted, "We are becoming more vulnerable to natural disasters because of the trends of our society rather than those of nature. In other words, we are placing more property in harm's way."

High-cost events, such as a devastating earthquake, simply are not factored into the way our society now plans and develops communities. In fact, economic incentives that would encourage more responsible ways to use land are actively denounced or, worse, stifled. The kinds of institutions and programs that have been established actually subsidize bad choices. For example, they offer low-interest rebuilding loans after an earthquake or help to rebuild beachfronts after a hurricane.

There is much in this terrestrial process that we now see reflected in the way we conduct activities in space. Satellite insurers underwrite bad satellite designs by charging only slightly higher rates (3.7 percent per year compared to 1.2 percent for "good" satellites) for poor or risky designs. They pay no attention to the solar activity cycle in establishing their rates. And once the stage has been set, it is difficult to change old habits. Like homeowners rebuilding property along receding coastlines after a hurricane, satellite owners insist on orbiting low-cost, inherently vulnerable satellites. Electrical utilities, meanwhile, forego investment in even the simplest mitigation technologies and prefer to view GIC problems as a local technical difficulty with a specific piece of equipment.

In the next ten years, the excitement of Cycle 23 will fall behind us, just as the heyday of Cycle 22 has now become a historical footnote. In 2006, we will find ourselves at the beginning of a new round of solar storminess. The great experiments of the LEO communication satellite networks will have run their inevitable courses, either as intermittent successes or as technological dinosaurs. Barring any major blackouts this time around, electrical utilities operating in a deregulated climate will debate on a region by region basis, whether the investment in GIC forecasting makes sense or not. We will also have weathered several years of round-the-clock occupancy of the International Space Station. This latter experience will teach us many lessons—some harsh—about what it really takes to deal with space weather events and what is needed to become permanent occupants of space. At some time, perhaps in the twenty-second century, our present concern and obsession with harmful space weather events will be a thing of the past. There is even the hope that the Sun may, unexpectedly, cease its stormy, cyclical behavior for a century or two as it has before. However, the last time the Sun became inactive in the 1600s, Europe experienced its famous mini ice age for several decades.

But, for now—for this year, for this cycle—we must remain vigilant even in the face of what seems like little cause for outright concern. For every pager that is temporarily silenced, others will work just fine. For every city that is darkened or loses air conditioning, a hundred others will experience just another ordinary day. Space weather is like a shark whose fin we see gliding across the interplanetary waters. We know it is there, but we haven't completely figured out what to do when it strikes.

Notes

The story of how scientists came to understand solar activity and its geophysical effects is a long and complicated one. Here are a few short essays that provide a bit more insight into some of the issues covered in this book. For more details visit the Astronomy Cafe web site at http://www.theastronomycafe.net

Chapter 3. "Hello: Is Anyone There?"

While Benjamin Franklin was flying a kite hoping to entice a lightning bolt into a jar, Sir William Watson in England sent another kind of electrical discharge from a battery down a wire some two miles long. It wouldn't have amounted to more than an odd laboratory curiosity if it hadn't been for a Frenchman named Lesage some twenty-five years later who found a rather odd application for it. He arranged a set of wires and batteries, one for each letter of the alphabet, and a distant reader could tell what letter was being sent by seeing which wire was charged or not. It was a comically strange way to send a message, but it was the first attempt at sending information that didn't use the centuries-old methods of smoke, mirrors, lanterns, or flags.

Marconi fully expected that radio broadcasting would be resilient to solar disturbances compared to telegraphy and telephony, because it used a very different medium to transmit its signals. While disturbances from the September 1909 Great Aurora were recorded worldwide in a variety of telegraph and telephone systems, he considered this storm and its impacts a lesson to be learned, not by wireless telegraphy, but by the competing communications technologies. In 1926, another Great Aurora lit up the skies, bringing this twenty-five-year sense of security to an abrupt end. International wireless tests with U.S. shortwave radio operators attempted to pick up stations in Wales, Argentina, and Peru. Electrical disturbances interfered with both broadcasting and telegraph services. Scientists blamed all of this on an unusually large sunspot visible on the Sun. Exactly one solar rotation later, on February 23, the same sunspot group was positioned as it had been for the January storm, and again problems erupted on the telegraph wires and in the ether. This time, shortwave radio reception of stations to the north of Ames, Iowa were blocked. Stations to the south came through clear as a bell. The third and final storm of this series arrived a week later from a different group of sunspots near the center of the Sun's disk. Again, voltage surges in the telephone lines were recorded, and shortwave reception only improved after the surges ended.

Chapter 4. Between a Rock and a Hard Place

On November 9, 1965, the largest blackout in history erupted in the United States in what became known as the Great Northeast Blackout. The event started at the Niagara generating station when a single transmission line tripped. Within 2.5 seconds, five other lines became overloaded and caused generators to become unstable and go off-line. Within four seconds, thirty million people in New York, Vermont, Massachusetts, and Connecticut lost power for up to thirteen hours. Later that day, President Lyndon Johnson directed the Federal Power Commission to investigate this failure. The full resources of the federal government, the Federal Bureau of Investigation, and the Department of Defense were called upon to support this investigation. There were many lessons learned from this nonspace weather blackout and today's electrical utility network is the result. The power grid of the 1960s had very few built-in safeguards that could have stifled this failure, it was also much less networked and interconnected. Paradoxically, Maine did not suffer the blackout because it was much less connected to the rest of the grid then than it is now.

More recently, on April 29, 1991, a transformer at the Maine Yankee Nuclear Plant catastrophically failed a few hours after a severe geomagnetic storm. This power plant has been closed since the mid-1990s because it is among the oldest nuclear power plants and it could no longer operate economically. Although the words *nuclear plant* and *transformer failure* appear in the same catastrophe, there was never any danger to the safety of the plant.

Engineers were eventually able to figure out that, for power lines, the only way to harden them against geomagnetically induced currents (GICs) was to build them over rock strata of the right conductivity. You could also make transformer cores forty times larger, increase your operating margins, and stop using the "Wye-type" open ground coupling scheme. Constructing transformers forty times larger is out of the question, because the transportation of these 100-ton devices is already at the limits of what can be accomplished with conventional trains and trucks. It is also not feasible to use another coupling scheme because Wye-type grounding is the least expensive, and any changes would cost many billions of dollars to implement. Because of a lack of genuine interest in such sporadic events, the recognition that they could be a significant new problem for large power grids came very slowly. Even by 1968 these types of disruptions were still so infrequent that most utilities could not muster more than a modest interest in them. None of the alterations that could reduce GIC impacts were practical options by 1995, so the only recourse was to attempt forecasting, if you felt so inclined.

In July 1998, electrical power transmission congestion in Wisconsin and Illinois blocked the transport of power from northern supplies to consumers in southern states who were sweltering in the heat. Power lines can only transport a fixed amount of power, and the essential transmission lines along the electrical superhighway were experiencing the equivalent of gridlock. Not enough volunteers could be found in the south to stop using their air conditioners, so the local electrical utilities had to go to local energy suppliers to purchase temporary "makeup" energy. Within a few hours, the price per megawatt soared from $20 to $7,000 and wiped out the yearly profits from several southern utilities. A similar problem occurred in 1999 during hot weather in the Midwest and Northeast. Deregulation has forced utilities into a wait-and-see mode where investments in infrastructure are postponed and new capacity is not planned. All of these factors work together to make even minor geomagnetic storms a potential "straw" that can break the back of regional electric power pools.

Blackouts and power interruptions of one kind or another are actually more common than you might suppose. Some last only fractions of a second, while others can last nearly a week. During a severe ice storm in 1999, seventy thousand people lost power for up to five days during a cold winter in the Washington, D.C. and Maryland region. New York City lost power for several days on a hot Wednesday in July 1977. Ten million people were without electricity, and two thousand were arrested for looting. In July–August 1996, ten states and six million people were without power for sixteen hours. The Electric Power Research Institute in Palo Alto, California has estimated that even short interruptions—sags—lasting less than half a second can cost U.S. companies $12 to $26 billion each year. Some paper companies lose $50,000 each time one of these dips comes along every week or so and stops five-ton spools of paper dead in their tracks, shredding hundreds of yards of paper and jamming presses.

The electrical power industry is slowly beginning to take GICs seriously because of several factors that have come to light in recent years. Power grid performance has become so optimized that any sources of inefficiency have become intolerable. GICs damage expensive equipment and generate VARs, which rob power companies of millions of dollars of potential revenue. Like a tax auditor, GICs can come at any time and affect hundreds of transformers in one stroke. Margins have also diminished so that there is less surplus power to cover emergency situations.

Developers of GIC monitors and forecasting technology are currently on the stump to sell their systems to power companies, but the process is hampered by power company managers who do not fully accept that GICs are a financial liability for them they can do something about. Reports of distant and sporadic blackouts (Quebec) have not fully impressed U.S. power managers because they do not affect their customers directly. In a $214-billion-a-year industry involving 700,000 miles of high-voltage power lines, losing a transformer every few years—or dealing with a solar storm-induced blackout against the hubris of other blackouts and sags—seems an unwarranted financial concern. Besides, although electrical utility customers reasonably expect their local power company to do all they can to keep the power running during an ice storm, solar storms are not a part of this expectation. They are a rare and "cosmic" event that most of us will pardon power companies from worrying about. But, sometimes, rare events can make a difference in tragic ways.

Chapter 6. They Call Them "Satellite Anomalies"

The Marecs-1 satellite, also suffered a complete failure on March 25, 1991. This satellite had a history of space environment problems. Its predecessor, Marecs-A launched in December 1981 had already been disabled ten years earlier by the strong electrical currents flowing during a week of intense auroral activity in February 1982.

Chapter 7. Business as Usual

The loss of *Intelsat 708* during a launch in the People's Republic of China triggered a congressional investigation on the role of commercial space insurance in technology transfer to the PRC. The August 1999 Cox Report was the outcome of this investigation, and it publicly revealed many of the details of how satellite insurance operates. To get insurance, a satellite owner selects an insurance broker who acts as an intermediary between the insurance underwriters and the satellite owner. The broker writes the policy, manages transactions, and settles claims. Brokers do not lose money in the event of an accident but are paid a commission on the basis of the size of the insurance package they write. The satellite owner prepares a technical document, giving a detailed assessment of the satellite and launch vehicle and any risks associated with the technology. This is presented to the broker, who then presents this to the various underwriters during the negotiation phase. This information is confidential and cannot be divulged to the public. Brokers and underwriters often retain their own staffs of independent technical experts, space scientists, and engineers to advise on the risk factors and to decide upon appropriate premium rates. The policy is then negotiated, with the broker serving as the intermediary between owner and underwriter. This can take up to three years prior to launch for major satellite systems. A 10–20 percent deposit is paid to the underwriters no later than thirty days before launch. Typically, the premiums are from 8–15 percent for the launch itself. In-orbit policies tend to be about 1.2 to 1.5 percent per year for a planned ten- to fifteen-year life span once a satellite survives its shakeout period.

According to Michael Vinter, vice president of International Space Brokers in Virginia, this period was once as short as one year but has now grown to as long as five years depending on the perceived riskiness of the satellite. If a satellite experiences environmental or technological problems in orbit during the initial shakeout period, the insurance premium paid by the satellite owner can jump to 3.5–3.7 percent for the duration of the satellite's lifetime. This is the only avenue that insurers have currently agreed upon to protect themselves against the possibility of a complete satellite failure.

Once an insurance policy is negotiated, the only way that an insurer can avoid paying out on the full cost of the satellite is in the event of war, nuclear detonation, confiscation, electromagnetic interference, or willful acts by the satellite owner that jeopardize the satellite. There is no provision for "acts of God" such as solar storms or other environmental problems. Insurers assume that if a satellite is sensitive to space weather effects, this will show up in the reliability of the satellite, which would then cause the insurer to invoke the higher premium rates during the remaining life of the satellite. Insurers, currently, do not pay any attention to the solar cycle.

Chapter 10. Through a Crystal Ball

The Great Aurora of September 9, 1859, lit up the skies around the world and caught astronomer Richard Carrington's eye just as he was about to end his observing session at the telescope. Carrington was an avid watcher of sunspots, and he had been watching a spectacular sunspot round the western limb of the Sun during the last few days. Within minutes, a powerful optical flare burst into light and then vanished. Meanwhile, miles away at the Kew Observatory outside London, the local magnetic field went haywire. This flare did much more than merely tilt compass needles and make a few astronomers sit upright. In France, telegraphic connections were disrupted as sparks literally flew from the long transmission lines. Huge auroras also blazed in the sky as far south as Hawaii, Cuba, and Chile. People spoke about this now long forgotten event much as we have obsessed about "killer asteroids" in recent years.

Despite the coincidence of flare and aurora, Carrington's observation was actually a fluke. Astronomers know that such brightenings visible to the eye through a telescope are literally a once-in-a-lifetime event and require especially titanic releases of energy on the Sun. For the next fifty years after Carrington's sighting, many careful studies were made of the solar surface and magnetic storm records, but no other sudden brightenings of the solar surface were ever seen again. It wasn't until the invention of the spectroheliograph and its successor, the visible light spectrohelioscope, between 1892 and 1910, that many more sudden brightenings were captured and their geomagnetic impacts could be properly assessed. Ultimately, the only proven way to anticipate solar flares, and the geomagnetic and ionospheric effects that might follow, is to watch the solar surface itself. Constantly.

Since the 1960s, solar physicists have known that sunspots with opposite polarity cores (called umbrae) within the same envelope (called penumbrae) were a potent spawning ground for flare activity. If a flare had been spotted near an active region, the odds were excellent that there would be more flares to follow from this same region over the course of the next few weeks. It didn't matter how big the sunspot group might be. What counted was how tangled up the magnetic field was in a small region of the solar surface. In the 1970s, new magnetic imaging technologies allowed flaring regions to be correlated with areas where strong shearing was occurring: magnetic fields with opposite polarities were trapped in regions in which gas motions were dynamically moving the magnetic fields around in very small parcels of gas. This seemed to be the crucial observational clue to anticipating when a flare is likely to breakout.

The BearAlert program eventually established an "eight-fold way" for evaluating whether conditions were ripe for a flare event or not. Current, official techniques used by NOAA's Space Environment Center use images of the entire Sun, rather than detailed studies of individual active regions, and tend to be accurate only about 25 percent of the time. The BearAlerts, with their much more detailed assessments of individual sunspot groups, scored correct predictions for M- and X-class flares about 72 percent of the time. What is also encouraging is that the method developed by Zirin and Marquette rarely misses the really big M-class flares that can do astronauts and satellites serious harm. The amount of lead time we have for solar flares has now expanded from literally a few minutes to several days. There is some indication, however, that a perfect record of correct calls may be forever out of reach. Solar activity, at the scales that trigger flare events, is largely a random process, just as the pattern of lightning strikes during a thunderstorm.

The number of geomagnetically disturbed days rises and falls with the sunspot cycle. The largest number seems to peak a year or so before, and a year or so after, sunspot maximum. The reason for this is not known. These disturbances seem to be more intense in the March–April and September–October periods as well. Here we think we understand this pattern a little better. The Earth's orbit is tilted five degrees to the equator of the Sun. This means that there will be two "seasons" during the Earth year, around the equinoxes, when the Earth passes through the equatorial plane of the Sun. The Sun also crosses the equatorial plane of the Earth at this time. Under these conditions, the southward-directed field of the Sun has its maximum strength, making it very powerful in stimulating geomagnetic storms. If you want the best chance of seeing a dramatic aurora, wait until sunspot maximum conditions prevail and visit northern latitudes during the March and September equinoxes.

As useful as Kp is, it does little to give you a meaningful advanced warning of what will soon be happening where you are located. Once you see the Kp index growing in size to become a major storm, the damage to your technology has already been done. Historical information about past storms tells the unhappy tale that, by the time you see Kp grow to the level of a medium-sized storm with Kp = 6, you have a roughly one in five chance it will continue to grow into a large storm with Kp = 7. You also have a roughly one in fifteen chance it will become a major storm with Kp = 9. It only takes a few hours for these kinds of changes to play themselves out. More troubling than this, geomagnetic conditions can look fairly normal for hours, then, within minutes, suddenly deteriorate into a severe storm. Despite its limitations for advanced warning, Kp is in many ways the only indicator that is readily available each day, so a variety of groups and industries find even this kind of information better than none at all: the electrical power industry for instance.

Although plasmas, fields, and currents form systems of staggering complexity, there are still consistent patterns of cause and effect that can be traced with considerable mathematical precision. There is nothing ad hoc about how a current of particles will generate a specific amount of magnetic field strength. It doesn't matter if the current is one ampere of electrons in a wire or a dilute five hundred thousand-ampere river of plasma orbiting the Earth. Maxwell's famous equations, combined with suitable "equations of motion," are in principle all that you need to describe the essential features of any "magneto-hydrodynamic" system such as the Earth and Sun. But, even with the theoretical game plan clearly defined, there is still a lot that is left unspecified. Theorists have a large number of mathematical choices to make in deciding which ingredients to keep and which to throw out. The more sources and interactions you add to your equations, the messier they become, and the harder it is to wrest a concrete mathematical prediction from them. High-quality data is the only looking glass that lets space scientists hit upon the right clues to guide them. Like learning how to dance, it is important to start with the correct foot forward, and only a careful study of Nature gives us the right choreography. Eventually, space scientists managed to win their way to a rather firm set of procedures for tackling questions about the Sun-Earth system. These "arrivals" were not in the form of some monolithic, single, comprehensive theory of how the whole shebang worked but a series of minor victories that comprised their own separate pieces in a larger puzzle.

For example, astronomers, armed with telescope and spectroscope, investigated the solar surface and pieced together the physical structure of the photosphere-chromosphere-corona region. They dissected the chemical compositions of the solar gases, measured their temperature, density, and speed, and crafted a working model of the solar atmosphere. They used powerful new "Zeeman-splitting" techniques to measure surface magnetic fields. With Maxwell's equations, the magnetic data helped theorists build models of the geometry of this field around sunspots and extend them deep into the corona. By 1960, a preliminary theory of why there is a sunspot cycle, and why sunspots occur, was hammered out by Eugene Parker at the University of Chicago and Horace Babcock at the Hale Observatories. Parker also went on to craft a groundbreaking theory, and mathematical description, of the solar wind as it leaves the coronal regions and flows throughout the solar system. Solar physics was, essentially, described by the complex mathematics of magneto-hydrodynamics. The particular phenomena we observed was "only" the working out by the Sun of specific mathematical solutions, driven by its complex convecting surface. What remained to be understood were the details of just how the solar magnetic field was generated, how the corona was heated, and why solar flares and other impulsive events get spawned. The missing link seemed to be the various gyrations of the magnetic field itself, but only new instruments in space would let scientists chase the magnetic forces down the rabbit's hole of decreasing size.

By the way, you should always keep in mind that things could be far worse for us than they are! For decades, astronomers have been studying stars that are close cousins to our Sun, a middle-aged G2-class star. At Mount Wilson Observatory, careful measurements of some of these stars show a distinct rise and fall in certain spectral lines that on our own Sun are indicators of solar activity. These stars also show periodic "sunspot cycles" with periods from a few years up to thirty years per cycle. Others show a constant level of activity, as our own Sun would have during the Maunder Minimum between 1610 and 1700. So solar activity is not unusual among the kinds of stars similar to our Sun. Rather alarming is that some kindred stars belt out super flares from time to time. In fact, according to Yale astronomer Bradley Schaefer, sunlike stars normally produce one of these superflares every century: "One of these cases I have is a star, S-Fornax, where for a 40-minute period it was seen to be three magnitudes brighter than usual." The power from the flare made the star appear nearly twenty times brighter than usual.

One of these superflares would be about ten thousand times more powerful than the solar storm that caused the 1989 Quebec blackout! According to Schaefer, portions of the surfaces of the outer ice moons of the solar system might be melted, much of the ozone layer would be destroyed, and the entire satellite fleet would be permanently disabled. It is believed that the reason the Sun doesn't have these flares is that it doesn't have a close companion star or planet that is magnetically active and able to tangle up our Sun's magnetic field.

Meanwhile, back at the Earth, the challenges were nearly as daunting. The shape of the Earth's magnetic field was eventually defined by numerous ground-level measurements and, with Maxwell's equations, extended thousands of miles into space. Although the general shape was still much like that of a simple bar magnet, there were noticeable lumps to it that followed geological changes in surface rock conductivity and subsurface irregularities reaching all the way to the core of the Earth itself. By the 1930s, physicists Sydney Chapman and Vincenzo Ferraro had mathematically described the impact that an "intermittent" solar wind would have upon the Earth's magnetic field. It was a staggering tour de force, linking together many separate geophysical systems and phenomena. The compression of the sunward side of the field would eventually lead to the amplification of a powerful ring of current flowing in the equatorial zone. Aurora had been studied meticulously since the nineteenth century and eventually gave up their quantum ghosts once the spectroscope was invented. Something was kicking the atmospheric atoms of oxygen and nitrogen so that they glowed in a handful of specific wavelengths of light. Through the rather contentious technical debates that began with Kristian Birkelund in 1896 and ended with Hannes Alfven in the 1950s, the general details of how aurora are produced came into clearer view. Some process in the distant geotail region was accelerating currents of electrons and protons along polar magnetic field lines. Within minutes, the currents dashed against the atmosphere and gave up their billions of watts of energy. There was, however, no detailed mathematical model that could recover all the specific shapes and forms so characteristic of these displays. This much was certain, however: we were living inside the equivalent of a TV picture tube, and the electron beams from the distant geotail region were drawing magical shapes on the phosphor screen of the sky.

The dawn of the Space Age brought with it an appreciation of most of the main ingredients to the complete geospace environment. All that seemed to be lacking in moving the frontier forward was more data to describe the geospace system in ever more detail. New rounds of complex equations needed to be fed still more detailed data to keep them in harmony with the real world. Space physics had reached a watershed moment where mathematically precise theories were sorely in need of specific types of data to help them further evolve. One small step along this way was to create a series of "average" models of the particles and fields in geospace.

NASA became a leader in developing and refining models of the Earth's environment through the Trapped Radiation Environment Modeling Program (TREMP) in preparation for the Apollo moon landings. The models combined the measurements made by dozens of satellites such as *Telstar* and *Explorer* and even instruments carried aboard the Gemini spacecraft. They didn't attempt to explain *why* the conditions were what they were, or how that got that way. Unlike the specific theories of the Sun-Earth system and its various components, TREMP program models, such as AE-8 and AP-8, were merely statistical averages of measured conditions in space and in different localities during solar maximum and solar minimum conditions only. They could not predict conditions that had not already been detected from the smoothed averages. The models did not include solar flares or other short-term and unpredictable events that can substantially increase accumulated radiation dosages. This was the best that could be done by the 1970s, and it is amazing that these models are still in wide use over thirty years later. Although they are adequate for designing satellite radiation shielding, they are useless for forecasting when the next storm will arrive. Some researchers don't even think they are all that useful for high-accuracy satellite shielding design.

Bibliography

ABC News, Reuters News Service 1999. The Sun's loaded gun: S-Shapes on surface foretell massive solar blasts, March 10. http://solar.physics.montana.edu/press/ABC/solarblasts990310.html

Adams, W. G. 1881. Magnetic disturbances, auroras, and earth currents. *Nature* 25:66.

Ainsworth, Diane. 1996. Briefing directly linking solar storms to disruptions on earth. JPL Briefing, February 15. http://www.calix.com/nasamail/Feb_1996/msg00008.html

Allen, Joe. 1990. 1989 solar activity and its consequences at Earth and in space. Seventh Symposium on Single Event Effects Conference, April 24–25, Los Angeles.

Allen, Joe, H. Saver, Lou Frank, and Patricia Reiff. 1989. Effects of the March 1989 solar activity. *Eos Transactions*, November 14, p. 1479.

American Journal of Science and Arts staff. Aurora borealis [of November 17, 1848]. *American Journal of Science and Arts* 12:442.

Anselmo, Joseph. 1997. Solar storm eyed as satellite killer. *Aviation Week and Space Technology*, January 27, p. 61.

Associated Press staff. 1999. Solar flare shock wave is photographed. *Washington Post*, April 10, A7.

Atherton, Tony. 1994. Cable firms scramble to get around hi-tech hijinx. *Ottawa Citizen*, January 22, p. G1.

Aviation Week and Space Technology staff. 1994. Anik E2 disabled. *Aviation Week and Space Technology*, January 31, p. 28.

——— 1998. The importance of spacecraft. *Aviation Week and Space Technology*, May 25, p. 17.

Bacon, John. 1999. Radiation Alert. *USA Today*, December 10, p. 3A.

Badhwar, G. and D. Robbins. 1996. Decay rate of the second radiation belt. *Advances in Space Research* 17:151.

Baker, Dan. 1998. Disturbed space environment may have been related to pager satellite failure. *EOS Transactions*, October 6, p. 477.

Baker, R. R., J. G. Mathew, and J. H. Kennaugh. 1983. Magnetic bones in human sinuses. *Science* 301:78–80.

Barrett, W. 1886. Earth-Currents and Aurora. *Nature* 34:408.

Barthez, Paul. 1998. *VM 525 Radiation Biology and Radiation Protection*. Ohio State University. http://www2.vet.ohio-state.edu/docs/clinsci/radiology/vm525/lecture6/lect6.html

Belian R. et al. 1992. High-Z energetic particles at geosynchronous orbit during the great solar proton event series of October 1989. *Journal of Geophysical Research* 97:16897.

Berger, Brian. 2000. NASA sees benefits from watching Sun. *Space News*, February 28, p. 3.

Bernard, Fredrick. 1837. Aurora Borealis of November 14, 1837. *American Journal of Science and Arts* 34:286.

Blakemore, Richard 1975. Magnetotactic bacteria. *Science* 190:377.

Blakemore, Richard, and Richard Frankel. 1981. Magnetic navigation in bacteria. *Scientific American* 245:58–65.

Bone, Neil. 1991. *The Aurora: Sun-Earth Connections*, p. 5. New York: Wiley.

——— 1996. *The Aurora: Sun-Earth Interactions*, p. 22. New York: Wiley.

Boss, Lewis. 1941. The aurora of September 18–19, 1941. *Popular Astronomy* 49:504.

Boston Globe staff. 1892. Tide of light: Telegraphic communication suffers some. *Boston Globe*, February 14, p. 2.

———1940. U.S. hit by magnetic storm: Wires, cable, radio crippled for hours. *Boston Globe*, March 25, p. 1.

——— 1946. Aurora borealis flash startles hub area. *Boston Globe*, July 27, p. 1.

——— 1972. Satellite trouble. *Boston Globe*, August 5, p. 2.

——— 1989. Quebec blackout halts power for 3 million. *Boston Globe*, March 14, p. 6.

——— 1997. Not much flare to this solar event, experts say. *Boston Globe*, April 10, p. A24.

Bougeret, J. L. 1999. Space weather: Final report from the French evaluation group on needs. *Centre National D'Etudes Spatiales* (July), p. 34.

Bowditch, Nathaniel. 1962. *American Practical Navigator*, pp. 23–24. Washington, D.C.: U.S. Government Printing Office.

Brausch, Jay. 1999. Spectacular aurora graces North Dakota [September 25, 1998]. *Astronomy* (March), p. 126.

Bredmeier, Kenneth. 2000. Helping area keep cool: Power companies say air conditioners will have plenty of electricity. [Bill Richardson quote.] *Washington Post*, June 10, 2000, p. E1.

Brekke, Pal. 1998. Solar eruptions and effects on the Earth's environment. May 18. http://www.uio.no/˜paalb/report/report.html

Broad, William. 1989. Space shuttle problem could cut flight short, *New York Times*, March 15, p. A16.

Burnett, Thane. 1994. Panic over Anik. *Toronto Sun*. January 21, p. 7.

Campbell, Wallace. 1978. Induction of auroral zone electric currents within the Alaska pipeline. *Pure and Applied Geophysics* 116:1143.

——— 1980. Observations of electric currents in the Alaskan oil pipeline. *Geophysical Journal of the Royal Astronomical Society* 61:437.

——— 1996. *Surveys in Geophysics* 8:239.

——— 1997. *Introduction to Geomagnetic Fields*, pp. 213–244. Cambridge: Cambridge University Press.

Chandler, David. 1999. Scientists read sun signs for danger. *Boston Globe*, March 10, p. A8.

Chester, Charles. 1848. On the electric telegraph of Prof. Morse. *Journal of American Science and Arts* 5:55.

Chianello, Joanne. 1994. Telsat's signal scrambled: Loss of power on E1 satellite will have long-term repercussions. *Ottawa Citizen*, March 30, p. E1.

Chicago Daily News. 1958. Aurora puts on display in northern sky. *Chicago Daily Tribune*, February 11, p. 4.

——— 1958. Helicopter on mercy mission crashes: 3 die. *Chicago Daily News*, February 10, p. 5.

Chicago Tribune staff. 1946. Chicagoans see sky alight with aurora display. *Chicago Tribune*, July 27, p. 5.

——— 1972. Solar flare causes static: magnetic storm. *Chicago Tribune*, August 3, p. 6.

——— 1972. Spectacular sky show caused by giant solar flare. *Chicago Tribune*, August 4, p. 3.

——— 1972. Jets nearly collide over Tokyo; 501 safe. *Chicago Tribune*, August 4, p. 2.

——— 1972. Exploding gases on sun start magnetic storms. *Chicago Tribune*, August 6, p. 12.

——— 1998. Talk about a communications breakdown [Kidnews]. *Chicago Tribune*, May 26, p. 3.

Conroy, Dave. 1997. *The Great Northeast Blackout of 1965*. Central Main Power Company. http://www.cmpco.com/aboutCMP/powersystem/blackout.html

Cosper, Amy. 1999. Ka-band questuions. *Satellite Communication*, July, p. 25.

Cowen, Ron. 1990. Cosmic radiation creates unfriendly skies. *Science News* 137:118.

Cox Report. 1999. *Commercial Space Insurance*, chapter 8. http://www.house.gov/coxreport/cont/gncont.html

Cucinotta, Francis. 1991. Radiation Risk Predictions for Space Station Freedom Orbits. NASA *Technical Paper* 8098.

Curran, Peggy. 1989. Hydro blames sun for power failure. *Montreal Gazette*, March 14, p. 1.

——— 1989. Hydro execs start post-mortem on huge blackout. *Montreal Gazette*, March 15, p. A3.

Curtis, Jan. 1999. Photography of major aurora seen in Alaska: 1996–2000. http://climate.gi.alaska.edu/Curtis/curtis.html

David, Leonard. 1999. Engineers still unable to explain NEAR's miss. *Space News*, June 14, p. 1.

Davidson, Art. 1990. *In the Wake of the Exxon Valdez*, pp. 8–19. San Francisco: Sierra Club.

DeCotis, Mark. 1997. AT&T's Telstar 401 satellite fails. *Florida Today: Space Online*. http://www.flatoday.com/space/explore/stories/1997/011397f.htm

de la Rive, A. 1854. On the cause of the aurora borealis. *American Journal of Science and Arts* 18:353.

de Selding, Peter. 1997. Insurers battle with satellite makers over quality control. *Space News*, April 28, p. 6.

——— 1997. Some satellite owners could be overpaying for insurance. *Space News*, April 28, p. 6.

——— 1997. Planned satellite systems minimize space debris. *Space News*, March 31, p. 1.

——— 1997. Insurers warn against false expectations: New underwriters unaware of risks. *Space News*, March 3, p. 3.

——— 1998. Iridium swaps some tests for launch speed. *Space News*, September 21, p. 3.

——— 1999. Satellite failures put big squeeze on underwriters. *Space News*, January 11, p. 1.

——— 1999. Iridium satellites please Motorola, despite losses. *Space News*, April 5, p. 1.

———1999. Vague satellite policies increase insurance claims. *Space News*, April 12, p. 1.

——— 1999. Insurers plan to raise premiums: Underwriters feel pressure of mounting satellite claims. *Space News*, May 15, p. 1.

——— 1999. Teledesic dream stalled. *Space News*, July 12, p. 1.

——— 1999. Wary investors avoiding satellite deals: Bankers vow to look at future plans skeptically. *Space News*, September 13, p. 1.

———1999. Hughes blames HS 601 glitches on design flaws. *Space News*, September 20, p. 23.

——— 1999. Solution to satellite defect still eludes Matra. *Space News*, September 20, p. 1.

——— 1999. ICO plans to cut costs, delay service: U.S. bankruptcy court oversees company's financial rehabilitation. *Space News*, October 25, p. 1.

——— 1999. Globalstar phone makers remain cautious. *Space News*, October 25, p. 1.

——— 1999. Immarsat approves internet satellites: Board wants system in orbit by 2004. *Space News*, December 13, p. 3.

——— 2000. Spacecraft losses make insurers wary. *Space News*, February 7, p. 1.

de Selding, Peter, and Sam Silverstein. 1999. Electronics advances boost small satellite capabilities. *Space News*, August 9, p. 1.

Discover staff. 1999. Shelter from the solar storm. *Discover*, October, p. 33.

Dietrich, William et al. 1942. Auroral burst supreme. *Sky and Telescope* 1:14.

Dooling, Dave. 1995. Stormy weather in space, *IEEE Spectrum* (June), p. 64.

——— 1999. Finding the smoking gun before it fires. *Space News*, March 9, p. 1. http://solar.physics.montana.edu/press/MSFC/spot_ast.htm

Duflot, Philippe-Alain. 1998. The viewpoint of the financial and insurance industries, Sixth Satel Conseil symposium, September 8–10, Maison de la Chimie, Paris, France. http://www.satelconseil.com/Sixth/symposium/conferen/session8/Conf_82.htm

Ebert, Bruce. 1999. Satellite communications: Instant infrastructure for the new millennium. *World Trade* (May), p. 34.

Eisele, Anne. 1997. NASA urged to pursue study of radiation effects. *Space News*, January 6, p. 6.

——— 1997. Soho data may enhance solar storm prediction. *Space News* September 1, p. 6.

——— 1998. EarlyBird blacks out in orbit. *Space News*, January 5, p. 1.

Electric Research. 1998. Geomagnetic Disturbance Monitoring: Operation Sunburst. http://www.electric-research.com/pq4.html

Emory University Radiation Safety Ofice. 1998. *Laboratory worker training manual*. http://www.cc.emory.edu/EHSO/r/train/train3.htm

Fahie, John 1974. *A history of electric telegrapy to the year 1837*. New York: Arno.

Fairbanks Daily News staff. 1989. Solar flare possibly the biggest ever recorded. *Fairbanks Daily News*, March 12, p. E10.

——— 1989. Solar storm sets night skies afire. *Fairbanks Daily News*, March 14, p. 3.

Ferester, Warren. 1997. Experts disagree on severity of next solar storm. *Space News*, January 20, p. 3.

——— 1997. Space weather monitoring faces funding woes. *Space News*, June 30, p. 6.

——— 1997. U.S. Scientists warn of rise in solar flares. *Space News*, December 1, p. 4.

——— 1998. NASA hunts for satellites to carry science payloads. *Space News*, April 6, p. 6.

——— 1998. Iridium taps long march to launch replacements. *Space News*, May 4, p. 1.

——— 1999. More in-orbit spares planned: Recent failures spur steps to guarantee service. *Space News*, June 28, p. 1.

Fink, Jim. 1997. Hale-Bopp through curtain aurora. *Astronomy* (September), p. 117.

Fisher, Lawrence. 1997. Failure of AT&T satellite will alter deal with Loral. *New York Times*, January 18, p. 36.

Flavelle, Dana and Donovan Vincent. 2 satellites break down within hours: Millions of viewers see TV screens go blank. *Toronto Star*, January 21, p. A1.

——— 1994. Communication satellite breakdown. *Toronto Star*, January 21, p. C1.

Foltman, Bob. Lights go out as U.S., Ireland Tie [Sports]. *Chicago Tribune*, June 7, p. 6.

Fortescue, Peter and John Stark. 1995. *Spacecraft Systems Engineering*. New York: Wiley.

Frazier, Kendrick. 1980. *Our Turbulent Sun*, pp. 12–19. Englewood Cliffs, N.J.: Prentice-Hall.

Friedman, Howard, Robert Becker, and Charles Bachman. 1963. Geomagnetic parameters and psychiatric hospital admissions. *Nature* 200:626.

Gifford, James. 1998. Risks you can live with. *Satellite Communications* (July), p. 6.

Green, Arthur and William Brown. 1997. Reducing the risk from geomagnetic hazards: On the watch for geomagnetic storms. United States Geological Service Fact Sheet [USGS177–97]. http://geohazards.cr.usgs.gov/factsheets/html_files/geomag/geomag.html

Hamer, Mick. 1981. Radiation agency notes the rising menace of radon. *New Scientist*, March 18, p. 31.

Hanlon, Michael. 1994. Sun's electrons beaned Anik 1 and 2. *Toronto Star*. January 29, p. A2.

Harwood, William. 1999. Shuttle's flat-panel future. *Washington Post*, May 3.

Henry, Joseph. 1847. On the induction of atmospheric electricity on the wires of the electrical telegraph. *Journal of American Science and Arts* 3:25.

Herrick, E. 1849. Aurora borealis of November 17, 1848. *American Journal of Science and Arts* 7:127, 7:293.

——— 1860. The great auroral exhibition of August 28th to September 4th, 1859. *American Journal of Science and Arts* 28:385–408, 29:92–97.

Herron, Edward. 1969. *Miracle of the Air Waves: A History of Radio*, p. 68. New York: Messner.

Highton, Edward. 1852. *The Electric Telegraph: Its History and Progress*, p. 144. London: J. Weale.

Hoffman, Steve. 1996. Enhancing power grid reliability. *EPRI Journal* (December).

Holden, Constance. 1999. Monarch Magnetism. *Science*, December 12.

Holliman, John. 1997. Sun ejection killed TV satellite. CNN.com, January 21. http://www.cnn.com/TECH/9701/21/cosmic.chaos

Horowitz, Carolyn. 1994. Here comes the sun. *Satellite Communications* (March), p. 10.

Houston Chronicle staff. The day the muzak died. *Houston Chronicle*, June 23, p. 1 [Business].

——— 1998. Radio station skirts busted satellite via 'Net; Planet shrinks. *Houston Chronicle*, May 25, p. 36.

Houston, Walter Scott. 1981. April's intense auroral display. *Sky and Telescope* 62:86.

International Atomic Energy Agency. 1998. *Radiation Safety*. http://www.iaea.or.at/worldatom/inforesource/other/radiation/radsafe.html

International Space and Terrestrial Physics Program. 1998. Coordinated observations of 1998 events including media coverage and multi-mission studies of CME events. April-May 1998. http://www-istp .gsfc.nasa.gov/istp/events

Jackson, Bernard and Paul Hicks. 1996. A CAT scan of the solar wind. http://www.sdsc.edu/GatherScatter/Gsfall96/jackson.html

Jaworowski, Zbigniew.1999. Radiation risk and ethics. *Physics Today* (September), p. 24.

Jayaraman, K. 1999. Sensor failure hurts India's weather effort. *Space News*, November 29. p. 1.

Jordan, Thomas and E. Stassinopoulos. 1989. Effective radiation reduction in space station and missions beyond the magnetosphere. *Advances in Space Research* 9:261.

Johnson, Jason. 1997. Power failure throws stadium for a loop. *Boston Herald*, January 13, p. 7.

Joselyn, JoAnn. 1995. Geomagnetic activity forecasting: The state of the art. *Reviews of Geophysics* 33:383.

Joselyn, JoAnn et al. 1996. "Panel Achieves Consensus Prediction of Solar Cycle 23. *EOS Transactions*, American Geophysical Union, May 20. http://www.sel.noaa.gov/info/Cycle23.html

Kallender, Paul. 1997. Japan's Adeos plagued by another problem. *Space News*, January 13, p. 1.

——— 1997. Failure of Adeos fuels debate on size of satellites. *Space News*, July 7.

——— 1998. ETS-7 problems unresolved. *Space News*, February 9, p. 1.

——— 1998. ETS-7 to begin docking tests. *Space News*, June 15, p. 3.

——— 1999. Japan developing satellite to warn of solar flares. *Space News*, May 3, p. 1.

Kallender, Paul and Sam Silverstein. 1999. Satellite failures spur power-amplifier developments. *Space News*, August 2.

Kappenman, John. 1998. Geomagnetic storms and impacts on power systems: Lessons learned from solar cycle 22 and outlook for solar cycle 23. August 1998. http://www.mpelectric.com/storms

——— 1997. Geomagnetic storm forecasts and the power industry. *EOS Transactions*, January 29, p. 37.

——— 1990. Bracing for the geomagnetic storms. *IEEE Spectrum* (March), pp. 27–32.

Kappenman, John and Vernon Albertson. 1990. Bracing for the geomagnetic storms. *IEEE Spectrum* (March), p. 27.

Kappenman, John and William Radasky. 1999. Learning to live in a dangerous solar system: Advanced geomagnetic storm forecasting technologies allow electric power industry to manage storm impacts. *Ohm Magazine* (Submitted).

Karmin, Craig. 1999. Tales of the tape: Telecom battle to control the sea beginning. *Business Today*, December 5, 1999.

Kirchvink, Jones. 1985 *Magnetite Biomineralization and Magnetoreception in Organisms*. New York: Plenum.

——— 1994. Rock magnetism linked to human brain magnetite. *EOS Transactions* 75:178.

Kluger, Jeffrey. 1999. Forecasting solar storms. *Time* (March), p. 88.

Kovalick, Tami 1999. The Coordinated Data Analysis web (CDAWEB) data archive interface. http://cdaweb.gsfc.nasa.gov/cdaweb/istp_public

Lang, Kenneth. 1995. *Sun, Earth and Sky*, p. 185. New York: Springer.

Larner, David. 1997. Sun flips bits in chips: Some computer crashes could be caused by neutron bombardment from Sun according to new evidence. *Electronic Times*, November 10, p. 1.

Lauriente, M. and J. Gaudet. 1994. Environmentally induced spacecraft anomalies. *Journal of Spacecraft and Rockets* (March-April), p. 153.

Leary, Warren. 1997. Researchers get first detailed look at magnetic cloud from sun. *New York Times*, January 23, p. A17.

Lefroy, J. 1852. Second report on observations of the aurora borealis, 1850–1851. *American Journal of Science and Arts* 14:155.

Leinbach, Harold. 1956. February sunspots and a red aurora. *Sky and Telescope* 15:329.

Letaw, J. et al. 1986. *Manned Mars Mission Working Papers*, part 2. NASA M-002.

Linn, Gene. 1999. Satellite failures rock insurers; shakeout, premium rise seen. *Journal of Commerce Online*, May 28. http://www.joc.com/issues/990528/i1nsur/e18979.htm

London Times staff. 1903. Telegraph disturbance in France. *London Times*, November 2, p. 6.

——— 1921. Sunspots and Earth storms. *London Times*, May 19, p. 7.

———1926. U.S. wireless reception experiments: partial success. *London Times*, January 28, p. 13.

Loral Space Systems. 1999. How we doubled satellite lifetimes in less than 8 years. *Space News*, February 22, p. 16.

Los Angeles Times staff. 1997. AT&T working to regain link with satellite: Telecom Telstar 401 is part of communications business to be sold to Loral for $712.5 million. *Los Angeles Times*, January 15, p. 1.

Lucid, Shannon. 1998. Six months on MIR. *Scientific American* (May), p. 51.

Ly, Phuong. 2000. Power failure disrupts national airport. *Washington Post*, p. B2.

McCaffery, Richard. 1997. TEMPO 2 Investigation Delays PAS-6 Launch. *Space News*, May 5, p. 1.

——— 1997. "Faulty Materials Blamed in Failure of Telstar 401. *Space News*, May 26, p. 1.

——— 1997. PAS-6 exhibits same design flaw as Loral's Tempo. *Space News*, October 20, p. 1.

——— 1998. Loral Reveals Limits of High-Power Satellites. *Space News*, March 9, p. 3.

——— 1998. Satellite makers use cheaper, faster approach. *Space News*, February 16, p. 15.

——— 1998. Satellite makers using cheaper, faster approach: Off-the-shelf parts allow smaller firms to compete. *Space News*, April 28, p. 1.

——— 1998. Globalstar, Iridium enjoy soaring market values. *Space News*, February 23, p. 1.

McIntosh, Patrick. 1972. August solar activity and its geophysical effects. *Sky and Telescope* 44:215.

Mansfield, Simon. 1998. Sandia to develop Intel rad-hard chips. *SpaceDaily Online*. http://www.spacer.com/spacecast/news/radiation-98c.html

Martignano. M. and R. Harboe-Sorenson. 1955. IBM ThinkPad 750C. *IEEE Transactions on Nuclear Science* 42:2004.

Maugh, Thomas. 1982. Magnetic navigation, an attractive possibility. *Science* 215:1492.

Maynard, Nelson. 1995. The National Space Weather Program. http://earth.agu.org/revgeophys/maynar01/node1.html

Mendell, David. 2000. Thursday outage not tied to weather Con Ed says. *Chicago Tribune*, June 10, p. 5.

Michener, James. 1982. *Space*, p. 489. New York: Random House.

Mills, Mike. 1998. Undersea cables carry growing rivers of data. *Washington Post*, March 9, p. E1.

Molinsky, Tom. 1999. Why Utilities FEAR GICs. XXVIth General Assembly, International University of Radio Science, August 1999. Paper E2.

Montgomery, John. 1998. Satellite loss claims double. *Space News*, September 22, p. 1.

——— 1997. The orbiting internet: Fiber in the sky. *Byte* (November), p. 58.

Montgomery, Monty. 1998. Aurora in northeastern Wisconsin. (photograph). *Astronomy* (March), p. 124.

Montreal Gazette staff. 1989. 11-hour power failure stopped the presses. *Montreal Gazette*. March 12, p. A3.

——— 1989. Hydros export clients lend a hand with power. *Montreal Gazette*. March 14, A3.

Morgan, Tom. 1999. *Janes Space Directory 1998–1999*. London: Butler and Tanner.

Mottelay, P. F. 1975. *The Bibliographical History of Electricity and Magnetism*, p. 1. New York: Arno.

Muir, Hazel. 2000. Animal magnetism: Do creatures use a chemical compass to find their way? *New Scientist*, June 10, p. 10.

National Academy of Science, 1998. Readiness for the upcoming solar maximum: Agency activities and related recommendations. http://www.nas.edu/ssb/maxch4.htm

NASA Facts. 1996. Our Sun: A look under the hood. http://spacescience.nasa.gov/pubs/oursun.htm

NASA, Johnson Space Center. 1998. Spaceflight Radiation Health Program at the Johnson Space Center. http://srag-nt.jsc.nasa.gov/docs/TM104782/techmemo.htm

NASA. 2000. Sun-Earth connections roadmap. http://www.lmsal.com/sec/roadmap/

NOAA Space Environment Services. 1998. Solar Proton Events list, 1976 to May 1998. http://umbra.nascom.nasa.gov/SEP/seps.html

NOAA National Geophysical Data Center. 1999. Geomagnetic Indices. ftp://ftp.ngdc.noaa.gov/STP/GEOMAGNETIC_DATA/INDICES/KP_AP

NOAA National Geophysical Data Center. 1999. Solar Proton Events list, 1990 to June 1999. ftp://ftp.ngdc.noaa.gov/STP/SOLAR_DATA/SAT_ENVIRONMENT/PARTICLES/p_events.html

National Research Council Space Studies Board. 1998. Space Weather: a research perspective. http://www.nas.edu/ssb/spwpt5nw.html

National Space Weather Program. 1999. Implementation Plan Chapter 1.2. http://www.ofcm.gov/homepage/text/pubs_linx.htm

Nature staff. 1938. The aurora of January 25. *Nature* 141:232.

New Scientist staff. 1989. Radiation agency notes the rising menace of radon. *New Scientist*. March 18, p. 31.

New Steel staff. 1998. Steelmakers zapped by power outages and higher electricity rates. *New Steel Journal* (August). http://www.newsteel.com/news/NW980801.htm

New York Times staff. 1882. A storm of electricity: Telegraph wires useless for several hours. *New York Times*, November 18, p. 1.

——— 1903. Northern lights display: Telegraph and cable lines suffer by electrical disturbance. *New York Times*, November 1, 1903, p. 1.

——— 1903. Electric phenomena in parts of Europe: telephone and street car services suspended in Switzerland. *New York Times*, November 2, p. 7.

——— 1903. Made the wires deadly. *New York Times*, November 1, p. 6.

——— 1907. Liner in collision. *New York Times*, February 10, p. 4.

——— 1917. Wires held up: Earth currents put them out of commission for hours. *New York Times*, August 9, p. 8.

——— 1937. Sunspots predict radio disturbances. *New York Times*, April 5, p. 15.

——— 1937. Magnetic storm worst in century. *New York Times*, April 29, p. 23.

——— 1937. Aurora borealis hits wire services:Current from magnetic pole causes record interference in Canada. *New York Times*, April 29, p. 23.

——— 1938. Aurora borealis startles Europe: People flee in fear, call firemen. *New York Times*, January 26, p. 25.

——— 1940. Sunspot tornado disrupts cables, phones and telegraphs for hours: Electrical disturbance plays havoc with the short-wave, 1,000,000 Easter messages, and police and press teletypes. *New York Times*, March 25, p. 1.

——— 1941. Northern lights soon: Unexpected sunspots may cause magnetic storms. *New York Times*. September 18, p. 1.

——— 1941. Aurora borealis gives city a show as sun spots disorganize radio. *New York Times*. September 19, p. 25.
——— 1941. Sunspots add some radio spice." *New York Times*, September 20, p. 41.
——— 1946. Sun spots block radio messages: Magnetic storm is sweeping the Earth due to intensify for 12 more days. *New York Times*, February 3, p. 26.
——— 1956. Rare aurora glimpsed: Northern lights with red arc seen in Alaska. *New York Times*, February 26, p. 44.
——— 1956. Suns raging storms photographed. *New York Times*, February 25, p. 1.
——— 1956. Six planes with 16 missionaries reported missing. *New York Times*. February 24, p. 1.
——— 1958. Sky brilliance among New England's finest shows in 30 years. *Boston Globe*, February 11, p. 27.
——— 1958. Aurora borealis keeps US radios disrupted. *New York Times*, February 12, p. 16.
——— 1958. One Explorer radio silent after 11 days. *New York Times*. February 13, p. 1.
——— 1989. Large solar flare erupts anew. *New York Times*, March 13, p. 1.
——— 1989. Helicopter crash kills fifteen in Arizona. *New York Times*, March 14, A22.
——— 1989. Five hundred on two trains reported killed by Soviet gas pipeline explosion. *New York Times*, June 5, p. 1.
——— 1989. At the soviet inferno: "Such suffering." *New York Times*, June 8, p. 1.
——— 1990. U.S. acts to assure safety of Alaska pipeline. *New York Times*, December 6, p. D19.
Nicastro, Anthony. 1981. Flare stars. *Astronomy* (June), p. 67.
North American Electric Reliability Council. 1998. *System conditions: 1998–2007 reliability assessment* (September). http://www.nerc.com/~ac/syscond.html
Odenwald, Sten. 1999. Solar storms: coming to a sky near you. *Washington Post*, March 10, p. H01.
——— 2000. Solar storms: The silent menace. *Sky and Telescope* (March), p. 51.
Office of the Federal Coordinator for Meterology. 1999. *National Space Weather Program*. http://www.ofcm.gov/homepage/text/pubs_linx/htm

Olmsted, Denison. 1837. Observations of the Aurora Borealis of January 25, 1837. *American Journal of Science and Arts* 32:176.

——— 1852. Great Aurora of February 19, 1852. *American Journal of Science and Arts* 13:426.

Orlando Sentinel staff. 1997. AT&T moves signals after satellite halts broadcast. *Orlando Sentinel*, January 12, p. A22.

Osetta, A. A. Favetto and E. Lopez. 1998. Currents induced by geomagnetic storms on buried pipelines as a cause of corrosion. *Journal of Applied Geophysics* 38:219.

Pansini, Anthony. 1996. *Guide to Electrical Power Distribution Systems.* Tulsa: Rennwell.

Peel, Quentin. 1989. Scale of Urals gas pipeline disaster stuns Soviet Union. *Financial Times of London*, June 6, p. 2.

Perera, Judith. 1989. March flares mark solar unrest. *New Scientist*, March 25.

Pirjola, Risto. 1999. Study exploring space weather risk to natural gas pipelines in Finland. *EOS Transactions*, July 27, p. 332.

Poppe, Barbara. 1999. *NOAA Space Weather Scales*. December 2, 1999. http://www.sec.noaa.gov/NOAAscales/index.html

Preece, W. 1873. The great aurora of February 4, 1872. *Journal of the Society of Telegraphic Engineers and Electricians* 2:114.

——— 1881. The electric storm of January 31st, 1881. *Journal of the Society of Telegraph Engineers and Electricians* 10:97–103.

——— 1894. Aurora of March 30, 1894 [strange noises on telegraph wires]. *Nature*, April 5, 49:539.

Quidlen, Terrey. 1999. [It's] Official: Geostationary Satellites Will Disappear. *Space News*, July 5, p. 3.

Reedy, Robert. 1997. Solar particle events and their radiation threats. *In Conference on the High Energy Radiation Background in Space*. IEEE Nuclear and Space Radiation Effects Conference, Snowmass, Colorado, July 22–23.

Reeves, Geoff. 1997. *Discover Magazine* (June), p. 36.

Refus, W. Carl. 1946. The beauty and mystery of the northern lights. *Sky and Telescope* 5:3.

Reuters Limited. 1997. Satellite failure hits Indian communications. October 5, 1997. cnn.com/TECH/9710/05/india.satellite.reut/index.com

——— 1998. $1.5 billion trans-Atlantic fiber optic cable project planned. http://www.techserver.com/newsroom/ntn/info/090298/info10_25345_noframes.html

Rinehart, Steve. 1989. Light show amazes, worries, lower 48 watchers. *Anchorage Daily News*, March 15, p. B11.

Roederer, Juan. 1995. Are magnetic storms hazardous to your health? *EOS Trans* 76:441.

Rosenfeld, Mary. 2000. Mary's little lambs [Fatima miracle]. *Washington Post*, June 10, 2000, p. C1.

Royce, Frederick. 1860. Auroral observations made at Washington, D.C. [telegrapher's near electrocution]. *American Journal of Science and Arts* 29:97.

Roylance, Frank. 1997. Tracking a solar storm from cradle to grave. *Baltimore Sun*, January 23, p. 1.

Ruzicka, Milan. 1998. Merrill Lunch bullish on space industry worldwide. *International Space Industries Report*, May 7, p. 1.

Schilling, Govert. 1999. Superflares from giant planets. *Science* 283:319.

Sky and Telescope staff. 1958. February's great multicolored aurora. *Sky and Telescope* 17:280.

——— 1991. *Sky and Telescope* 82:329, 82:216.

Sarnoff, David. 1936. The future of radio and public interest, convenience and necessity. *RCA Review* 1:5–12.

Sarton, G. 1927. *Introduction to the History of Science*, p. 410. Baltimore: William and Wilkins.

Satellite Communications staff. 1997. Uplink: Telstar 401 failure. *Satellite Communications* (March), p. 8.

Satellite News staff. 1997. AT&Ts Telstar 401 communications satellite declared dead: Impact on Skynet sale to Loral uncertain. *Satellite News*, January 20, p. 1.

——— 1998. Panamsat faces financial hit on Galaxy IV due to limited insurance. *Satellite News*, May 25, p. 1.

Savage, Candace. 1994. *Aurora: The Mysterious Northern Lights*, pp. 37–38. San Francisco: Sierra Club.

Savage, Don. 1997. Solar mystery nears solution with data from SOHO spacecraft. NASA *Press Release*, November 5, 1997, no. 97–256.

Sawyer, Kathy. 1999. Scientists find way to predict solar storms. *Washington Post*, March 10, 1999, p. A2.

Schenk, K. 1999. Coronal Mass Ejections list from SOHO/LASCO, January 1996 to December 1999. http://lasco-www.nrl.navy.mil/cmelist.html

Scott, William. 1997. Operators place high value on space weather forecasting. *Aviation Week and Space Technology*, September 1, p. 54.

Shea, M. D. Smart, J. Allen, and D. Wilkinson. 1992. Spacecraft problems in association with episodes of intense solar activity and related terrestrial phenomena during March 1991. *IEEE Transactions on Nuclear Science* 39:1754.

Showstack, Randy. 1998. Upcoming solar maximum could affect wide range of technologies. *EOS Transactions* 79:601.

Sawyer, Kathy. 1997. Earth takes a one-two punch from a solar magnetic cloud. *Washington Post*, January 23, p. 1.

Seattle Times staff. 1997. Solar storm sends particle wave toward Earth–Satellite detects eruption on sun: Spacecraft, power grids should be safe. *Seattle Times*, April 9, p. 1.

Sedgwick, W. T. and H. W. Tyler. 1939. A *Short History of Science*, p. 219. New York: MacMillian.

Seife, Charles. 1998. ISS Titanic. *New Scientist*, November 14, p. 38.

——— 1999. Thank our lucky stars. *New Scientist*, January 9, p. 15.

Shipman, Harry. 1989. *Humans in Space*, pp. 113–118. New York: Plenum.

Sheldon, Robert. 1998. On the physical origin and prediction of killer electron storms. *Journal of Geophysical Research* (Submitted). Summary at: http://cspar181.uah.edu/RbS/GRL5/grl_may.html; reply to referees at: http://cspar181.uah.edu/RbS/GRL5/referee1a.html

Sietzen, Frank. 1998. Protecting satellites from long-term radiation exposure. *SpaceDaily*, August 21. http://www.spacer.com/spacecast/news /radiation-98a.html

——— 1998. Radiation belts affect satellites. *SpaceDaily*, December 9. http://www.spacer.com

Silverstein, Sam. 1998. Iridium spares dwindle as start date nears. *Space News*, May 18, p. 3.

——— 1998. Teledesic shakes up team: Motorola plans fastest satellite construction ever. *Space News*, May 25, p. 1.

——— 1998. Familiar problem leaves Galaxy 7 with no backup. *Space News*, June 29, p. 1.

——— 1998. PAS-5 joins PanAmSat's list of troubled satellites. *Space News*, July 6, 1998, p. 20.

——— 1998. String of HS601 failures put Hughes on hot seat. *Space News*, July 13, p. 1.

——— 1998. More satellite failures threaten introduction of Iridium service. *Space News*, July 27, p. 1.

——— 1998. Reasons for failure lost with Galaxy 4. *Space News*, August 17, p. 1.

——— 1998. Galaxy loss puts squeeze on PanAmSat. *Space News*, August 31, p. 3.

——— 1998. Insurers beleaguered by rash of failures in '98. *Space News*, August 31. p. 1.

——— 1998. Staiano expects slew of on-orbit failures. *Space News*, September 14, p. 1.

——— 1998. Satellite makers, insurers weather rocky '98. *Space News*, December 7, p. 16.

——— 1999. Iridium officials insist venture will survive. *Space News*, July 26, p. 3.

——— 1999. Mororola gives Iridium deadline: New investor must be found by February 15, 2000. *Space News*, December 20, p. 3.

——— 2000. Pressure intensifies on satellite makers. *Space News*, January 17, p. 1.

Simonsen, Lisa. 1991. Radiation protection for human missions to the moon and mars. NASA *Technical Paper* 3079.

Singer, Jeremy. 2000. Fiber optic boom complicates NSA satellite strategy. *Space News*, June 12, p. 1.

Smith, Bruce. 1998. Motorola begins to work on Teledesic design requirements. *Aviation Week and Space Technology*, June 1, p. 25.

——— 1998. Additional Iridium failures drop stock. *International Space Industries Report*, August 3.

Stein, Keith. 1998. Hughes hit hard by satellite failures. *International Space Industry Report*, July 20, p. 1.

——— 1998. Additional Iridium failures drop stock. *International Space Industries Report*, August 3, p. 1.

——— 1998. Teledesic's first satellite not functioning as planned. *International Space Industry Report*, May 7, p. 1.

Steinle, Helmut. 1999. Max Planck Institute press release on Equator-S outage, May 1. http://www.mpe-garching.mpg.de/www_plas/EQS/eq-s-news.html

Smith, Gar. 1996. Climate change melts US power grid. *Earth Island Journal* (Fall). http://www.earthisland.org/journal/f96–22.html

Space News staff. 2000. Solar flares disrupt two commercial satellites. June 19, p. 2.

Taylor, John. 1998. Satellite woes to cost DTN $5.8 million. *Omaha World-Herald*, June 11, p. 22sf.

ThinkQuest, 1998. Map of Bleigh Reef and Exxon Valdez path. http://library.advanced.org/10867/spill/maps/tanker_lanes.jpg

Thomsen, Michell and Joe Borovsky. Reaction of magnetosphere to Jan 10–11 event, as seen by 3 MPA instruments at geosynchronous orbit. www-istp.gsfc.nasa.gov/istp/cloud_jan97/lanl-jan10–11.html

Toronto Star staff. 1989. Huge storms on Sun linked to blackout that crippled Quebec. *Toronto Star*, March 13, p. A3.

——— 1989. Carelessness cited in Soviet disaster. *Toronto Star*, June 6, p. A17.

——— 1994. Telesat explanation disputed. *Toronto Star*, January 25, A10.

Townsend, L., J. Nealy, J. Wilson, and W. Atwell. 1988. Large solar flare radiation shielding requirements for manned interplanetary missions. *Journal of Spacecraft* 26:126.

Tribble, Alan. 1995. *The Space Environment: Implications for Spacecraft Design*, pp. 138–165. Princeton: Princeton University Press.

——— 1995. *The Space Environment*, pp. 157–158. Princeton: Princeton University Press.

Uehling, Mark. 2000. Arsenal of the Sun. *Popular Science* (February), p. 52.

USA Today staff. Satellite failure reveals our high-tech vulnerability. *USA Today*, May 26, p. 13A.

Vallance Jones. A. 1992. Historical review of great auroras [from 34AD to 1989]. *Canadian Journal of Physics* 70:479.

Vampola, A. 1994. Analysis of environmentally induced spacecraft anomalies. *Journal of Spacecraft and Rocketry* (March-April), p. 154.

van der Vink, G. 1998. Why the United States is becoming more vulnerable to natural disasters. *EOS Transactions*, November 3, p. 533.

Verrengia, Joseph. 1995. Boulder scientist's solar reports helped allies plan D-day invasion. *Denver Rocky Mountain News*, September 27, p. 27A.

Vestner, Charlie. 1998. The sky's the limit: In the 21st century satellites will connect the globe. *Individual Investor* (June), pp. 56–64.

Vinter, Michael. 1999. International Space Brokers. http://www.isbworld.com/gateway.html

Wald, Mathew. 1997. Storm on sun viewed by spacecraft a million miles from earth. *New York Times*, April 10, A22.

Walker, Matt. 1998. Flirting with disaster: The ever-shrinking microchip is increasingly vulnerable to an invisible enemy. *New Scientist*, November 7, p. 4.

Washington Post staff. 1998. Cracking the fiber optic club. *Washington Post*. September 7, p. 20 [Business].

——— 1998. Digital flub: Bank computers take a hit. *Washington Post*. September 7, p. 5 [Business].

Webb, Joseph. 1999. Telecommunications history timeline. http://www.webbconsult.com/timeline.html

Weyland, M., F. Cucionotta, and A. Hardy. 1991. Analyses of risks associated with radiation exposure from past major solar particle events. NASA *Technical Paper* 3137.

White Sands Missle Range, Public Affairs Office, Trinity Site. 1998. *Radiation at Ground Zero.* http://www.wsmr.army.mil

Whiting, Kathryn and Edward Kovalcik. 1993. *AIAA Aerospace Design Engineers Guide*. 3d ed. Washington, D.C.: American Institute of Aeronautics and Astronautics.

Wilkinson, Daniel. 1991 TDRS-1 Single Event Upsets and the effect of the space environment. *IEEE Transactions on Nuclear Science* 38:1708.

Will, George 1999. Astronomy's Answer. *Washington Post*, January 30, A15.

Williams, Martyn. 1998. PanAmSat reported satellite anomalies last week. CNNfn, May 20, 1998. http://europe.cnnfn.com/digitaljam/mewsbytes/112244.html

Williamson, S. J., L. Kaufman and D. Brenner. 1977. Magnetic fields of the human brain. *U.S. Naval Research Review*, October 1–18.

Willis, H. Lee. 1997. *Power Distribution Planning Reference Book*. New York: Marcel Dekker.

Woodcock, Gordon. 1986. *Space Stations and Platforms*, pp. 60–68. Florida: Orbit.

World Book. 1963. *Year Book: Reviewing Events in 1962*, p. 461. Chicago: Field Enterprises Education.

Wrenn, Gordon. 1995. Conclusive evidence for internal dielectric charging anomalies on geosynchronous communication satellites. *Journal of Spacecraft and Rockets* 32:514.

Yee, Andrew. 1998. USGS Reports geomagnetic storm in progress. April 26, 1998 storm report. http://www-istp.gsfc.nasa.gov/scripts/Guarda/USDG_on_ms-8–26–98.html

Zirin, Harold and William Marquette. 1991. BearAlert: A successful flare prediction system. *Solar Physics* 131:149. http://www.bbso.njit.edu/BearAlert/bearalert_paper.html

Figure and Plate Credits

Figure 1.1: courtesy the Defense Meteorological Satellite Program and NOAA, National Geophysical Data Center.

Figure 4.1: courtesy the North American Electric Reliability Council.

Figure 6.1: courtesy Daniel Wilkinson, NOAA/NGDC.

Figure 9.1: data courtesy NOAA Space Environment Center.

Figure 9.2: CME data courtesy the SOHO/LASCO consortium. SOHO is a product of the international cooperation between ESA and NASA. Geomagnetic data and GOES flare data courtesy NOAA/NGDC.

Plate 1: courtesy NASA, STS-39.

Plate 2: NASA, TRACE mission.

Plate 3: courtesy SOHO/LASCO/NRL.

Plate 4: courtesy John Kappenman, Metatech Corporation.

Plate 5: courtesy SOHO/MIDI consortium. SOHO is a product of the international cooperation between ESA and NASA.

Plate 6: solar images courtesy SOHO/EIT consortium. SOHO is a product of the international cooperation between ESA and NASA.

Plate 7: ionosphere images courtesy National Geophysical Data Center, NOAA.

Plate 8: courtesy the NASA, IMAGE Satellite Team.

Plate 9: courtesy U.S. Air Force Space Environment Modeling Program.

Plate 10: courtesy John Kappenman, Metatech Corporation.

Plate 11: courtesy Dr. Bernard Jackson, University of California Santa Barbara.

All other figure and plate material courtesy the author.

Index